CLIMATE CHANGE AND ITS CAUSES, EFFECTS AND PREDICTION

CLIMATE CHANGE

ENVIRONMENTAL AND ECONOMIC EFFECTS

CLIMATE CHANGE AND ITS CAUSES, EFFECTS AND PREDICTION

Additional books and e-books in this series can be found on Nova's website under the Series tab.

Climate Change and its Causes, Effects and Prediction

Climate Change

Environmental and Economic Effects

Fukui Ayaka
Editor

snova
New York

Library of Congress Cataloging-in-Publication Data

ISBN: 978-1-53619-239-1

Published by Nova Science Publishers, Inc. † New York

CONTENTS

PREFACE

Climate change and extreme weather events present risks to our communities and communities around the world. Experts have warned that climate change would lead to more intense storms, extended droughts, longer wildfire seasons that burn hotter and cover larger areas, greater seasonal temperature extremes, melting of glaciers and ice sheets, and rising sea level. This book discusses ways to help communities better adapt to these changes.

Chapter 1 - This is an edited, reformatted and augmented version of Hearing before the Subcommittee on Environment and Climate Change, of the Committee on Energy and Commerce, House of Representatives, One Hundred Sixteenth Congress, First Session, Serial No. 116–1, dated February 6, 2019.

Chapter 2 - This is an edited, reformatted and augmented version of Hearing Before the Committee on Science, Space, and Technology, House of Representatives, One Hundred sixteenth Congress, First Session, Serial No. 116–35, dated July 11, 2019.

In: Climate Change
Editor: Fukui Ayaka

ISBN: 978-1-53619-239-1

Chapter 1

TIME FOR ACTION: ADDRESSING THE ENVIRONMENTAL AND ECONOMIC EFFECTS OF CLIMATE CHANGE*

Committee on Energy and Commerce

Wednesday, February 6, 2019

House of Representatives, Subcommittee on Environment and Climate Change, Committee on Energy and Commerce, Washington, DC.

The subcommittee met, pursuant to call, at 10:02 a.m., in the John D. Dingell Room 2123, Rayburn House Office Building, Hon. Paul Tonko (chairman of the subcommittee) presiding.

Members present: Representatives Tonko, Clarke, Peters, Barraga´ n, McEachin, Blunt Rochester, Soto, DeGette, Schakowsky, Matsui, McNerney, Ruiz, Pallone (ex officio), Shimkus (subcommittee ranking

---

* This is an edited, reformatted and augmented version of Hearing before the Subcommittee on Environment and Climate Change, of the Committee on Energy and Commerce, House of Representatives, One Hundred Sixteenth Congress, First Session, Serial No. 116–1, dated February 6, 2019.

member), Rodgers, McKinley, Johnson, Long, Flores, Mullin, Carter, Duncan, and Walden (ex officio).

Also present: Representatives Castor and Sarbanes.

Staff present: Jeffrey C. Carroll, Staff Director; Adam Fischer, Policy Analyst; Jean Fruci, Energy and Environment Policy Advisor; Tiffany Guarascio, Deputy Staff Director; Caitlin Haberman, Professional Staff Member; Rick Kessler, Senior Advisor and Staff Director, Energy and Environment; Brendan Larkin, Policy Coordinator; Dustin J. Maghamfar, Air and Climate Counsel; Tim Robinson, Chief Counsel; Mike Bloomquist, Minority Staff Director; Adam Buckalew, Minority Director of Coalitions and Deputy Chief Counsel, Health; Jerry Couri, Minority Deputy Chief Counsel, Environment; Jordan Davis, Minority Senior Advisor; Caleb Graff, Minority Professional Staff Member, Health; Peter Kielty, Minority General Counsel; Bijan Koohmaraie, Minority Counsel, CPAC; Ryan Long, Minority Deputy Staff Director; Mary Martin, Minority Chief Counsel, Energy and Environment; Brandon Mooney, Minority Deputy Chief Counsel, Energy; Brannon Rains, Minority Staff Assistant; Zack Roday, Minority Director of Communications; Peter Spencer; Minority Senior Professional Staff Member, Energy.

Mr. TONKO. Good morning, everyone, and welcome to the Subcommittee on Environment and Climate Change's first hearing of the year. Now that the gavel has been found, we can move forward.

Let me before I make my comments thank Chairman—former Chairman, always Chairman perhaps—John Shimkus for his great work in leading this subcommittee. I think we had an outstanding track record. And I enjoyed the years that he served as chair and I as ranking member. It is a pleasure to have served with you and now to continue to serve with you.

I welcome all the colleagues of this subcommittee to this first hearing and to service through this subcommittee. And in general I think we have a lot of business ahead of us but I look forward to a great, spirited debate on all of these issues and bipartisan response to the solutions that we will develop.

The subcommittee now comes to order. I recognize myself for 5 minutes for an opening statement.

Opening Statement of Hon. Paul Tonko, a Representative in Congress from the State of New York

In 1957, when I was the impressionable age of 8, Earth entered the Space Age with the launch of the Sputnik satellite by the Soviet Union. People around the world stopped what they were doing and looked the heavens. Nothing after that would ever be the same. Americans leapt into action, training to become scientists and engineers in droves. I was one of them.

And I see that same motivation, wonder, and drive in many of the people today who are working and advocating to transform our economy to one that is cleaner, safer, and more just. They are advancing clean energy technologies, designing the infrastructure of the future that will help communities endure, and rethinking every industry we have ever known.

It goes by many different names: Sandy, Harvey, Maria, Katrina, Campfire. But there is no question we have reached a new generation's Sputnik moment. How we respond to this threat and the opportunities it offers will indeed shape American lives for generations. In the 1960s our Government and our Nation's best rose to the Sputnik challenge by sending a person to the moon. Today our course remains unclear.

How our committee responds at this inflection point will define our Nation for the next half-century and beyond. Will we rise to this challenge and tackle our most complex problems? Will we continue to be the world leader in science, engineering, and technology innovation? Will we make our country and our planet better for future generations?

These questions are at the heart of our work here today. In 1961, when President Kennedy promised to put a person on the moon by the end of the decade, what would have been the consequences of failure? Loss of scientific discovery? Damage to America's reputation? Ultimately it would have been remembered as another missed deadline, or failed call to action, or broken promise from a politician.

With climate change, the cost of failure is existential. Failure to launch this next moonshot will result in deaths, devastation, and irreversible damage to our communities, our economy, and our environment. This is not an exaggeration. It is the assured outcome if we should fail.

But America is a nation of pioneers and problem solvers. This climate challenge is not beyond us. Time is running out but it is not gone. Some of our colleagues may protest the cost of climate protection. And our constituents are already paying a heavy price after each and every hurricane, wildfire, and flood. Investing in solutions and resilience today will help manage and limit those risks and serve as a foundation for job creation, healthier communities, and economic opportunity. But let's be clear: There is no path forward more costly than for us to do nothing.

Today we will hear from an expert panel to help us better understand those costs, along with possible solutions that Congress should consider. Dr. Brenda Ekwurzel coauthored the Fourth National Climate Assessment and can explain climate threats our Nation is facing.

Mike Williams can discuss job opportunities that will come from a clean energy transition, including from building more resilient infrastructure to adapt to new climate realities.

Reverend Leo Woodberry can tell us the importance of a transition that is equitable. We must address historic environmental injustices and ensure that benefits of a green transition are shared across every community.

Rick Duke can discuss a range of potential policy and technology solutions for climate mitigation, many of which are cost-competitive and proven to work.

In the decade since Congress last considered comprehensive climate legislation, green technologies have become more affordable and more effective. Today there are viable decarbonization pathways for many sectors of our economy that will enable our Nation and the world to achieve emissions reduction targets. Congress can give the certainty, price signals, and resources needed to achieve these goals.

In 1961, we chose to go to the moon. Today we must make another choice. Will we have the clarity of mind and conscience to choose to address climate change with the urgency that scientists say is necessary? I say yes. Chairman Pallone says yes. Every Member on this side of the aisle says yes. And we are willing to work with the legions of Americans, countless businesses, local, State, and foreign governments, our U.S. Department of Defense, and our colleagues here on the other side of the aisle, and anyone else with ideas that can solve this crisis.

To my friends across the aisle, I implore you, now is the time to join us. We want to work together, but inaction is no longer an option. We must act on climate.

These issues were not always partisan. Our parties came together to pass the Clean Air Act and its amendments. And as a credit to Mr. Shimkus' leadership, this subcommittee found ways to work together to solve other seemingly intractable, multi-decade stalemates. We have proven we can find common ground and we can get things done. We want to find solutions that work for all communities and all Americans, and we will not be deterred.

We have science-based targets that we cannot afford to miss. The very real and urgent threat of climate change is not just the issue of the day, it is the issue of our time, the challenge of our time, the opportunity of our time. And I hope the hearings held by this subcommittee will help us find a path, a path forward where we can seize this opportunity.

With that, I yield back.

[The prepared statement of Mr. Tonko follows:]

PREPARED STATEMENT OF HON. PAUL TONKO

In 1957, when I was at the impressionable age of 8, Earth entered the space age with the launch of the Sputnik satellite by the Soviet Union.

People around the world stopped what they were doing and looked to the heavens. Nothing after that would ever be the same. Americans leapt into action, training to become scientists and engineers in droves. I was one of them.

And I see that same motivation, wonder, and drive in many of the people today who are working and advocating to transform our economy to one that is cleaner, safer, and more just.

They are advancing clean energy technologies, designing the infrastructure of the future that will help communities endure, and rethinking every industry we have ever known.

It goes by many different names: Sandy, Harvey, Maria, Katrina, Camp Fire. But there is no question we have reached a new generation's Sputnik moment. How we respond to this threat, and the opportunities it offers, will shape American lives for generations.

In the 1960s, our Government and our Nation's best rose to the Sputnik challenge by sending a person to the moon. Today, our course remains unclear.

How our committee responds at this inflection point will define our Nation for the next half-century and beyond.

Will we rise to this challenge and tackle our most complex problems? Will we continue to be the world leader in science, engineering, and technology innovation? Will we make our country and planet better for future generations? These questions are at the heart of our work here today.

In 1961, when President Kennedy promised to put a man on the moon by the end of the decade, what would have been the consequences of failure? Loss of scientific discovery? Damage to America's reputation? Ultimately, it would have been remembered as another missed deadline, or failed call to action, or broken promise from a politician.

With climate change, the cost of failure is existential. Failure to launch this next moonshot will result in deaths, devastation, and irreversible damage to our communities, our economy, and our environment.

This is not an exaggeration. It is the assured outcome if we should fail.

But America is a nation of pioneers and problem solvers. This climate challenge is not beyond us. Time is running out, but it is not gone.

Some of our colleagues may protest the costs of climate protection, but our constituents are already paying a heavy price after every hurricane, wildfire, and flood.

Investing in solutions and resilience today will help manage and limit those risks, and serve as a foundation for job creation, healthier communities, and economic opportunity.

But let's be clear, there is no path forward more costly than for us to do nothing.

Today we will hear from an expert panel to help us better understand those costs, along with possible solutions that Congress should consider.

Dr. Brenda Ekwurzel coauthored the Fourth National Climate Assessment and can explain climate threats our Nation is facing.

Mike Williams can discuss job opportunities that will come from a clean energy transition, including from building more resilient infrastructure to adapt to new climate realities.

Rev. Leo Woodberry can tell us the importance of a transition that is equitable. We must address historic environmental injustices and ensure that benefits of a green transition are shared across every community.

Rick Duke can discuss a range of potential policy and technology solutions for climate mitigation, many of which are cost competitive and proven to work.

In the decade since Congress last considered comprehensive climate legislation, clean technologies have become more affordable and effective. Today there are viable decarbonization pathways for many sectors of our economy that will enable our Nation and the world to achieve emissions reduction targets.

Congress can give the certainty, price signals, and resources needed to achieve these goals.

In 1961, we chose to go to the moon. Today, we must make another choice. Will we have the clarity of mind and conscience to choose to address climate change with the urgency that scientists say is necessary?

I say yes. Chairman Pallone says yes. Every Member on this side says yes. And we are willing to work with the legions of Americans, countless businesses, local, State, and foreign governments, our U.S. Department of Defense, and anyone else with ideas that can solve this crisis.

To my friends across the aisle, I implore you: join us! We want to work together, but inaction is no longer an option. We must act on climate.

These issues were not always partisan. Our parties came together to pass the Clean Air Act and its amendments. And as a credit to Mr. Shimkus' leadership, this subcommittee found ways to work together to solve other seemingly intractable, multi-decade stalemates. We have proven we can find common ground and get things done.

We want to find solutions that work for all communities and all Americans, and we will not be deterred. We have science-based targets that we cannot afford to miss.

The very real and urgent threat of climate change is not just the issue of the day. It is the issue of our time. The challenge of our time. The opportunity of our time. And I hope the hearings held by this subcommittee will help us find a path forward where we can seize this opportunity. I yield back.

Mr. TONKO. And the Chair now recognizes Mr. Shimkus, ranking—excuse me, Republican leader of the Subcommittee on Environment and Climate Change, for 5 minutes for his opening statement.

OPENING STATEMENT OF HON. JOHN SHIMKUS, A REPRESENTATIVE IN CONGRESS FROM THE STATE OF ILLINOIS

Mr. SHIMKUS. First of all, let me congratulate you, Mr. Chairman. And thank you for the kind words. I am truly touched by those.

We have had some policy differences over the past 6 years. We also enjoyed, as you identified, some significant bipartisan policy achievements during my chairmanship, in no small part because of the thoughtful work that you brought to the panel as a Democrat leader, and your very competent staff. I believe this subcommittee will be served by your leadership.

Today's hearing ticks off a topic that will be challenging but not impossible to work through in a bipartisan manner. We all agree that extreme weather events and climate change presents risks to our communities and communities around the world. While we agree these risks should be addressed, we may disagree about what to do. If we are to reach an agreement on this issue, I believe we must look openly and broadly at potential solutions.

Many climate policy advocates have been suggesting for years that if you agree climate change is real, then command and control policy prescriptions are the only way to address this problem. If you question these expensive solutions, you must not accept the problem.

That is a false choice. And the amped-up partisan rhetoric it generates severely inhibits a full look at potential, practical policies that not only help reduce carbon dioxide emissions, but also ensure our Nation and its communities can grow and prosper.

Recent projections by the International Energy Agency show that fossil energy, even with all existing and announced policies implemented, will likely be the dominant form of energy in our world system through 2040, and likely beyond. Wind and solar energy will serve a larger portion of electricity generation across the world and in the United States according to this data, but fossil energy and nuclear energy, a technology regrettably frowned upon by many climate policy advocates, will remain dominant.

While future innovation could substantially change these projections, the stubborn route is that U.S. and global energy systems necessary for societies to develop, grow, trade, and prosper depend upon affordable and abundant energy and mobility. Policies that artificially raise the costs or availability of energy threaten to undermine this fundamental fact, which helps explain the 30-year failure of international climate agreements to significantly reduce global emissions, although the United States seems to be doing better than most of the countries that are in agreement.

No nation seeking to improve the lives of its citizens will accept energy or transportation constraints, and neither should the United States if we want to maintain a robust economy, economic growth, and remain globally competitive for future generations.

We could have a fuller conversation about accelerating the transformation to cleaner technologies if we accept that proposing topdown Government requirements to rapidly decarbonize the U.S. and global economies may not be the most realistic way to address the climate change problem.

We should be open to the fact that wealth transfer schemes suggested in the radical policies like the Green New Deal may not be the best path to community prosperity and preparedness.

And we should be willing to accept that affordable and abundant energy is a key ingredient for economic development and growth. After all, economic growth and economic resources, coupled with sound planning, infrastructure, and governance, increase local capabilities to minimize impacts of future extreme events.

These are realities we should explore today and in future hearings if we want to develop sound environmental and energy policies to address climate risk. We should also focus on the ingredients behind the exceptional achievements of American know-how in energy, in technology and innovation that has led to world-leading prosperity, and making sure we can continue to foster these advances in other technology.

The American shale revolution transformed our Nation's economic competitiveness and is driving cleaner electricity generation because of old-fashioned innovation, entrepreneurship, regulatory certain private capital, not bigger Government mandates. And let me also mention private property rights on these areas. Let's apply these lessons more broadly.

Mr. Chairman, there are different approaches to dealing with climate change. Let's focus on solutions that work for the American public.

And with that, Mr. Chairman, I yield back my time. [The prepared statement of Mr. Shimkus follows:]

PREPARED STATEMENT OF HON. JOHN SHIMKUS

First, let me congratulate you Mr. Chairman. While you and I had some policy differences over the past 6 years, we also enjoyed some significant bipartisan policy achievements during my chairmanship—in no small part because of the thoughtful work you brought to the panel as Democrat leader.

I believe this subcommittee will be well served with your leadership.

Today's hearing kicks off a topic that will be challenging, but not impossible, to work through in a bipartisan manner. We all agree that extreme weather events and climate change present risks to our communities-and communities around the world.

While we agree these risks should be addressed, we may disagree about what to do. If we are to reach an agreement on this issue, I believe we must look more openly and broadly at potential solutions.

Many climate policy advocates have been suggesting for years that, if you agree climate change is real, then command-and-control policy prescriptions are the only way to address the problem. If you question these expensive solutions, you must not accept the problem.

This is a false choice. And the amped up partisan rhetoric it generates severely inhibits a full look at potential, practical policies that not only help reduce carbon dioxide emissions but also ensure our Nation and its communities can grow and prosper.

Recent projections by the International Energy Agency show that fossil energy, even with all existing and announced policies implemented, will remain the dominant form of energy in our global systems through 2040, and likely beyond.

Wind and solar energy will serve a larger portion of electricity generation across the World and in the United States, according to this data, but fossil energy and nuclear energy—a technology regrettably frowned upon by many climate policy advocates—will remain dominant.

While future innovation could substantially change these projections, the stubborn reality is, the U.S. and global energy systems necessary for societies to develop, grow, trade, and prosper depend upon affordable (and abundant) energy and mobility.

Policies that artificially raise the cost or availability of energy threaten to undermine this fundamental fact, which helps explain the 30-year failure of international climate agreements to significantly reduce global emissions (although the United States seems to be doing better than most other nations).

No nation seeking to improve the lives of its citizens will accept energy or transportation constraints, and neither should the United States if we want to maintain robust economic growth and remain globally competitive for future generations.

We could have a fuller conversation about accelerating the transformation to cleaner technologies if we accept that proposing top-down Government requirements to rapidly decarbonize the U.S. and global economies may not be the most realistic way to address the climate change problem.

We should be open to the fact that wealth transfer schemes, suggested in radical policies like the Green New Deal, may not be the best path to community prosperity and preparedness.

And we should be willing to accept that affordable (and abundant) energy is a key ingredient for economic development and growth. Afterall, economic growth and economic resources, coupled with sound planning, infrastructure, and governance, increase local capabilities to minimize impacts of future extreme events.

These are realities we should explore today and in future hearings if we want to develop sound environmental and energy policies to address climate risks.

We should also focus on the ingredients behind the exceptional achievements of American know-how in energy, in technology, and in innovation that has led to world-leading prosperity—and make sure we can continue to foster these advances in other technologies.

The American shale revolution transformed our Nation's economic competitiveness and is driving cleaner electricity generation because of old-fashioned innovation, entrepreneurship, regulatory certainty, and private capital—not big Government mandates. Let's apply these lessons more broadly.

Mr. Chairman, there are different approaches to dealing with climate change. Let's focus on solutions that work for the American public.

Mr. TONKO. The gentleman yields back. And thank you, Mr. Shimkus.

The Chair now recognizes Mr. Pallone, chairman of the full committee, for 5 minutes for his opening statement.

Mr. Pallone.

OPENING STATEMENT OF HON. FRANK PALLONE, JR., A REPRESENTATIVE IN CONGRESS FROM THE STATE OF NEW JERSEY

Mr. PALLONE. Thank you, Mr. Tonko, Chairman Tonko.

Today's hearing on climate change is long overdue. We are feeling its effects now, and the influence of unchecked climate change is becoming more obvious every year. Experts have warned us for a long time that climate change would lead to more intense storms, extended droughts, longer wildfire seasons that burn hotter and cover larger areas, greater seasonal temperature extremes, melting of glaciers and ice sheets, and rising sea level.

The predictions have proven true. And these scientific experts warn us that, as greenhouse gas pollution continues to grow, climate change effects will intensify as the planet warms to levels that people have not experienced any time in human history.

Unfortunately, we are currently going in the wrong direction with respect to greenhouse gas pollution. The Fourth National Climate Assessment of the International Panel on Climate Change's recent report made clear that if we do not aggressively cut emissions now, we will jeopardize public health and safety, as well as our economic and national security.

The science on climate change is indisputable. And I do want to thank—I listened to Mr. Shimkus' opening remarks, and I noticed that he basically said that he agrees that there is a major impact from climate change, suggested that innovation was certainly one of the ways that we deal with it. So, again, I want to say that I know that in the past we were never able to have a hearing on climate change when the Republicans were in the majority, but I am glad to see that our ranking member is saying that it's something that has to be dealt with and is real.

I don't think that we need to debate the scientific facts. Instead, we must focus on solutions to the problems and must act now to avoid the most catastrophic consequences associated with climate change. The good news is that we already know the solutions. There are untapped opportunities to expand the use of renewable energy and to become more efficient with all the resources and energy we use. With focused investment and innovation, we can help transform industries and economic sectors that will find meaningful emission reductions more challenging.

Meanwhile, States, local government, and individual businesses are moving forward to reduce emissions to meet our obligations under the Paris Agreement. And it is now time for the Federal Government to step up and help them in these efforts and spur further action in communities across the country.

I know there are those who believe we can't address this problem because the costs are too high. But the costs of not acting are far higher and a lot more painful. In 2017, the U.S. experienced 16 natural disasters with costs totaling $360 billion. This past year, disasters again cost over $100 billion. The dollar figures are concerning, but the real tragedy is the loss of life and destruction of homes, businesses, and communities when these events occur.

And tremendous, sustained efforts are required for communities to recover and rebuild. And I saw this firsthand in the aftermath of Superstorm Sandy in my district. Events disappear from the headlines in a matter of weeks, but the work to rebuild and recover takes years. And it is still going on in my district. Many people have not been able to return to their homes. Many businesses have not.

We simply cannot afford to delay any longer. And we must discuss ways to help communities better adapt to the changes that we are already seeing. We need to modernize and upgrade our infrastructure to ensure vital services like water, sewer, electricity, telecommunications, and transportation are more resilient.

And here, Mr. Shimkus, in particular, I think that we can work together with the Republicans. And this important work would not only make our communities safer and better prepared for extreme weather events, but it will also provide good-paying jobs and the modern, flexible infrastructure that will better support a robust economy in the future.

We want to find innovative solutions that will help strengthen our economy by creating jobs in industries that will begin to repair the disparities found in so many vulnerable communities. And it is precisely those front-line communities that experience the worst effects of climate change and natural disasters and that are the least able to recover from them. Again, I saw it in my own district where some of the most vulnerable communities economically are the ones that still have not recovered.

I think we can do better. We must do better. And these communities need to be engaged in the process of designing adaptation and mitigation measures to reduce pollution.

So as we move forward, we hope to have our Republican colleagues as partners in these efforts. Certainly what has been said by Mr. Shimkus today gives me hope. The devastating effects of unchecked climate change do not know partisan or political boundaries. They effect us all. And I hope we will be able to find common ground and work together on solutions.

And the U.S. has always been a global leader in science, technology, and industry. And our leadership on climate action and global transformation to a low-carbon economy is leading now. This hearing is the start of our efforts to maintain U.S. leadership and to put us on the path to a low-carbon and more prosperous future.

And if I can say something, Chairman Tonko, I know that this has always been something that you cared so much about and worked on even when you were in the State legislature. So we are glad that you are the chairman. Thank you.

[The prepared statement of Mr. Pallone follows:]

PREPARED STATEMENT OF HON. FRANK PALLONE, JR.

Today's hearing on climate change is long overdue. We are feeling its effects now, and the influence of unchecked climate change is becoming more obvious every year. Experts have warned us for a long time that climate change would lead to more intense storms, extended droughts, longer wildfire seasons that burn hotter and cover larger areas, greater seasonal temperature extremes, melting of glaciers and ice sheets, and rising sea level. Their predictions have proven true. And, these scientific experts warn us that as greenhouse gas pollution continues to grow, climate change effects will intensify as the planet warms to levels that people have not experienced any time in human history.

Unfortunately, we are currently going in the wrong direction with respect to greenhouse gas pollution. The Fourth National Climate Assessment and the International Panel on Climate Change's recent report make clear that if we do not aggressively cut emissions now, we will jeopardize public health and safety, as well as our economic and national security.

The science on climate change is indisputable. We are not going to waste any time debating the scientific facts. Instead, we must focus on solutions to the problem. We must act now to avoid the most catastrophic consequences associated with climate change.

The good news is that we already know the solutions to this challenge. There are untapped opportunities to expand the use of renewable energy and to become more efficient with all the sources of energy we use. With focused investment and innovation, we can also help transform industries and economic sectors that will find meaningful emission reductions more challenging.

Meanwhile, States, local government and individual businesses are moving forward to reduce emissions to meet our obligations under the Paris Agreement. It's now time for the Federal Government to step up and help them in these efforts and spur further action in communities across the country.

I know there are those who believe we cannot address this problem because the costs are too high. But, the costs of not acting are far higher and more painful. In 2017, the U.S. experienced 16 natural disasters with costs totaling $360 billion. This past year disasters again cost over $100 billion. The dollar figures are concerning, but the real tragedy is the loss of life and destruction of homes, businesses, and communities when these events occur. Tremendous, sustained efforts are required for communities to recover and rebuild. I saw this first-hand in the aftermath of Superstorm Sandy in my district. Events disappear from the headlines in a matter of weeks, but the work to rebuild and recover takes years.

We simply cannot afford to delay any longer, and we must discuss ways to help communities better adapt to the changes that we're already seeing. We need to modernize and upgrade our infrastructure to ensure. vital services like water, sewer, electricity, telecommunications, and transportation are more resilient. This important work will not only make our communities safer and better prepared for extreme weather events, but it will also provide good paying jobs, and the modern, flexible infrastructure that will better support a robust economy in the future.

We want to find innovative solutions that will help strengthen our economy by creating new jobs and industries and that will begin to repair the disparities found in so many vulnerable communities. It is precisely these "front line" communities that experience the worst effects of climate change and natural disasters and that are the least able to recover from them. We can do better. We must do better. And, these communities need to be engaged in the process of designing adaptation and mitigation measures to reduce pollution.

As we move forward, we hope to have our Republican colleagues as partners in these efforts. The devastating effects of unchecked climate change—do not know partisan or political boundaries. They affect all of us. I hope we will be able to find common ground and work together on solutions.

We cannot transform our economy and society overnight, but every journey starts with a single step. The U.S. always has been a global leader in science, technology, and industry. And, our leadership on climate action and a global transformation to a low carbon economy is needed now. This hearing is the start of our effort to maintain U.S. leadership and to put us on the path to a low-carbon—and more prosperous—future.

I thank the witnesses for participating in this important hearing. I look forward to your testimony today and to working with you to address the climate challenge before us.

I yield back.

Mr. TONKO. Thank you, Mr. Chair. The gentleman yields back. And, Chairman Pallone, I appreciate your comments.

The Chair now recognizes Mr. Walden, the Republican leader of the full committee, for 5 minutes for his opening statement.

OPENING STATEMENT OF HON. GREG WALDEN, A REPRESENTATIVE IN CONGRESS FROM THE STATE OF OREGON

Mr. WALDEN. Well, thank you, Mr. Chairman. Again, congratulations on finding the gavel and using the gavel. We are delighted to work with you. And thanks for holding this hearing on climate change.

It is no secret the Energy and Commerce Committee has the jurisdiction, the ability to find a bipartisan path forward to tackle this important issue that confronts not only our Nation but also the world. As you know, I spoke out early and forcefully, Mr. Chairman, about the unnecessary effort by Speaker Pelosi to create yet a separate select committee which lacks any legislative authority. Our able Members will certainly serve on that panel. It is as redundant as the last one she created more than a decade ago.

With all this activity, it is important to highlight a few fundamentals at the onset. Climate change is real. The need to protect the environment is real. The need to foster a strong U.S. economy and grow American jobs is also real. And the need to prepare our communities for the future is real. Republicans on this committee are ready, willing, and able to have serious solutions-oriented discussions about how to address and balance these considerations.

For instance, we believe that a longer conversation about the Democrats' Green New Deal is necessary. We have heard about general tenets of the plan for the U.S., such as all-renewable electricity generation by 2030, all-zero-emission passenger vehicles in just 11 years, a Federal job guarantee, a living wage guarantee, but we obviously have some concerns about the potential adverse economic employment impacts of these measures.

At least one analysis has estimated that going to a 100 percent renewable energy in the U.S. could cost a minimum of $5.7 trillion—trillion—dollars. It sounds like a huge sum for consumers and taxpayers to foot.

The Republicans are focused on solutions that prioritize adaptation, innovation, and conservation. Just as America led the world in energy development, which reduced carbon emissions, we want America's innovators to develop the next technologies that will improve the environment and create jobs here at home. We want to help the environment for our children, and grandchildren, and their children. We also want the people who live in our districts in this country today, right now, to have jobs and to be able to provide for their families.

These are not mutually exclusive principles. And I believe, Mr. Chairman, working together we can develop the public policies to achieve these goals.

As the Republican leader of the committee, I will work to promote a better policy vision for the environment, one which supports and accelerates continued technological advances in energy and environmental practices to improve our quality of life. It ensures a sound regulatory environment where people have the confidence to invest their money to innovate and to create American jobs, one that improves information needed to understand future impacts and provide resources to communities to adapt and to prepare for these impacts, one that promotes America workforce development and training in energy-related industries, and one that recognizes the importance of open and competitive markets in the role the United States plays as the world's leading energy producer, innovator, and exporter of advanced technologies.

Indeed, Republicans have a track record of supporting policies that protect the environment and ensure energy access. For example, in the last Congress we supported legislation to promote zeroemissions nuclear energy, and renewable energy including hydropower. Hydropower has great success as a clean energy source across the country, and especially in my district and my State, where 40 percent of our energy comes from hydropower.

Legislation we passed into law in the last Congress will stream-line the permitting process for closed-loop pump hydropower projects. We have such a project in the permitting process in my district that would power up to 600,000 Oregon homes in a closed loop hydropower process.

We also advanced legislation to promote energy efficiency, grid modernization, energy storage, natural gas, a more resilient electric grid, carbon capture and utilization, and better forest management to address wildfires and limit their air quality impacts. This is what happens after a fire. This is called post-fire wildlife habitat right here. It is nothing but ash and destruction of the habitat.

Oregonians choke on smoke every summer from wildfires that burn across our poorly managed Federal forests, filling our skies with ash and polluting our airsheds with carbon dioxide, among other pollutants. Managing our forests not only reduces the risk of these catastrophic fires, but the Intergovernmental Panel on Climate Change say that sustainably managing our forests would create the longest sustained carbon mitigation benefit. So there is work we could do there.

And the numbers show that our policies are working. In 2017, U.S. carbon emissions were the lowest they have been since 1992, and are projected to remain steady in upcoming years, more than 10 percent below 2005 levels. Unfortunately, the Green New Deal ignores many of these important elements of our energy strategy and makes it more difficult to reach our shared environmental goals.

We look forward to hearing from our witnesses today on these topics, especially Mr. Powell from ClearPath, which has promoted clean energy, advanced nuclear, and carbon capture, and Mr. Worthington of the U.S. Energy Association, which advocated for a diverse energy mix within the United States and the importance of energy access and affordability around the globe.

So, when it comes to climate change, Mr. Chairman, Republicans are focused on solutions. That is why we back sensible, realistic, effective policies to tackle climate change. What we are deeply concerned about are plans we believe will harm consumers and cost American jobs and drive up our costs and not result in the kinds of goals we want to achieve mutually.

So thank you for having the hearing. I yield back the balance of my time.

[The prepared statement of Mr. Walden follows:]

PREPARED STATEMENT OF HON. GREG WALDEN

Thank you, Mr. Chairman, for holding this hearing on climate change. It is no secret that the Energy and Commerce Committee has the jurisdiction and ability to find a bipartisan path forward to tackle this important issue that confronts not just our Nation, but the world. As you know, I spoke out early and forcefully about the unnecessary effort by Speaker Pelosi to create a separate, select committee which lacks any legislative authority. While able Members will serve on this panel, it is as redundant as the last one she created more than a decade ago.

With all this activity, it is important to highlight a few fundamentals at the onset. Climate change is real. The need to protect the environment is real. The need to foster a strong U.S. economy and grow American jobs is real. And the need to prepare our communities for the future is real. The Republicans on this committee are ready and willing to have serious, solutions-oriented discussions about how to address and balance these considerations.

For instance, we believe that a longer conversation about the Democrats' Green New Deal is needed. We have heard about general tenets of the plan for the U.S.— such as all renewable electricity generation by 2030, all zero-emission passenger vehicles in just 11 years, a Federal job guarantee, and a living wage guarantee. We have serious concerns about the potential adverse economic and employment impacts of these types of measures. At least one analysis has estimated that going to 100 percent renewable energy in the U.S. could cost a minimum of $5.7 trillion— that sounds like a huge cost for consumers and taxpayers to foot.

Republicans are focused on solutions that prioritize adaptation, innovation, and conservation. Just as America led the world in energy development that has reduced carbon emissions, we want America's innovators to develop the next technologies that will improve the environment and create jobs here at home.

We want a healthy environment for our children, grandchildren, and their children. But we also want the people who live in our districts and in this country today, right now, to have jobs and to be able to provide for their families. These are not mutually exclusive principles. Working together we can develop the public policies to achieve these goals.

As the Republican leader on the committee, I will work to promote a better policy vision for the environment, one which:

- Supports and accelerates continued technological advances in energy and environmental practices to improve our quality of life;
- Ensures a sound regulatory environment, where people have the confidence to invest their money to innovate and create American jobs;
- Improves information needed to understand future impacts and provides resources to communities to adapt and prepare for those impacts;
- Promotes American workforce development and training in energy-related industries; and,

- Recognizes the importance of open and competitive markets; and the role the United States plays as the world's leading energy producer, innovator, and exporter of advanced technologies.

Indeed, Republicans have a track record of supporting policies that protect the environment and ensure energy access. For example, last Congress we supported legislation to promote zero-emissions nuclear energy, and renewable energy including hydropower. Hydropower has great success as a clean energy source in my Oregon district and generates approximately 40 percent of the electricity in my State. Legislation we passed into law last Congress will streamline the permitting process for closed-loop pumped hydropower projects. One such project in my district aims to generate enough power for 600,000 homes in southern Oregon.

We also advanced legislation to promote energy efficiency, grid modernization, energy storage, natural gas, a more resilient electric grid, carbon capture and utilization, and better forest management to address wildfires and limit their air quality impacts.

Oregonians choke on smoke every summer from wildfires that burn across our poorly managed Federal forests, filling our skies with ash and polluting our airsheds with carbon dioxide. Managing our forests not only reduces the risk of these catastrophic fires, but the Intergovernmental Panel on Climate Change says that sustainably managing our forests will create the longest sustained carbon mitigation benefit.

And the numbers show that our policies are working—in 2017, U.S. carbon emissions were the lowest they have been since 1992, and they are projected to remain steady in upcoming years, more than 10 percent below 2005 levels.

Unfortunately, the Green New Deal ignores many of these important elements of our energy strategy, and makes it more difficult to reach our shared environmental goals. I look forward to hearing from our witnesses today on these topics, particularly Mr. Powell from ClearPath, which has promoted clean energy, advanced nuclear and carbon capture, and Mr. Worthington of the U.S. Energy Association, which has advocated for a diverse energy mix within the United States, and the importance of energy access and affordability around the globe.

When it comes to climate change, Republicans are focused on solutions. That's why we back sensible, realistic, and effective policies to tackle climate change.

What we are deeply concerned about are the Democratic plans we believe will harm American consumers and American jobs by driving up costs and pushing jobs overseas where environmental laws are far more lax. We can do better than old policies rooted only in over-regulation, excessive-taxation, and economic stagnation.

Thank you, Chairman, and I yield back.

Mr. TONKO. Thank you, Representative Walden. And the gentleman yields back.

As chair, I remind Members that, pursuant to committee rules, all Members' written opening statements shall be made part of the record.

I now introduce our witnesses for today's hearing. And let me thank each and every one of you for sharing your time and offering input on this very important topic. We do appreciate your participation.

So we have from my left to right Dr. Brenda Ekwurzel, Director of Climate Science, Union of Concerned Scientists.

Next to her is Mr. Rich Powell, executive director of ClearPath.

Then we have Mr. Rick Duke, principal of Gigaton Strategies.

Then Reverend Leo Woodberry, Justice First Tour, Kingdom Living Temple Church.

Then we have Mr. Barry K. Worthington, executive director of United States Energy Association.

And then finally, Mr. Michael Williams, deputy director of BlueGreen Alliance.

We as a committee want to thank our witnesses for joining us today. We look forward to your testimony. At this time, the Chair will now recognize each witness for 5 minutes to provide his or her opening statement.

Before we begin I would like to explain the lighting system. In front of our witnesses is a series of lights. The lights will initially be green at the start of your opening statement. The light will turn yellow when you have 1 minute left. Please begin to wrap up your testimony at that point. The light will turn red when your time expires.

So, with that, Dr. Brenda Ekwurzel, again welcome. You are recognized for 5 minutes.

STATEMENTS OF BRENDA EKWURZEL, PHD, DIRECTOR OF CLIMATE SCIENCE, UNION OF CONCERNED SCIENTISTS; RICHARD J. POWELL, EXECUTIVE DIRECTOR, CLEARPATH; RICHARD D. DUKE, PRINCIPAL, GIGATON STRATEGIES; REVEREND LEO WOODBERRY, JUSTICE FIRST CAMPAIGN, KINGDOM LIVING TEMPLE CHURCH AND NEW ALPHA COMMUNITY DEVELOPMENT CORP.; BARRY WORTHINGTON, EXECUTIVE DIRECTOR, UNITED STATES ENERGY ASSOCIATION; AND MICHAEL WILLIAMS, DEPUTY DIRECTOR, BLUEGREEN ALLIANCE

Statement of Brenda Ekwurzel

Dr. EKWURZEL. Thank you, Chairman Tonko, Ranking Member Shimkus, and for the opening statements by Chairman Pallone and Ranking Member Walden, and the committee for providing me the opportunity to testify here before you today.

I am Director of Climate Science at the Union of Concerned Scientists, and I also had the privilege of serving as one of the coauthors of the Fourth National Climate Assessment released in November. Before I share with you the advances in our understanding from these latest assessments, I want to turn to a recent example of the high cost of climate change.

During the recent outbreak of extreme cold weather that gripped large parts of the Nation, a University of Iowa student and a University of

Vermont student were counted among at least 21 people who perished from consequences likely from the dangerous wind chill. Although it may seem counterintuitive, recent studies indicate that climate can cause unusually cold temperatures at midlatitudes by disrupting the normal winter season polar vortex in the stratosphere.

A good analogy to this disruption is a weak seal on a freezer door that periodically allows frigid air to flood into the room while warmer air rushes into the freezer. At the end of January, similarly, a cold blast spilled out of the Polar Regions and into the Midwest and expanded through to the eastern U.S., breaking wind chill records across. Yet Alaska experienced above-freezing temperatures and rain falling on snow, forcing the cancellation of middistance dog sled races that contestants use to compete for the long-distance races, the Iditarod.

Evidence is growing that warmer-than-normal periods in the Arctic are associated with a greater chance for extreme winter weather in the eastern United States. This deadly cold snap is just a recent example of the changing nature of extreme events that scientists are studying. One goal is to provide earlier warning so local officials have more time to take precautionary measures and improve safety.

Climate assessment provides the public and policymakers the most advanced warnings through summary and evaluation of the latest science. I will briefly share with you some findings with you today from the Intergovernmental Panel on Climate Change Special Report on Global Warming of 1.5 degrees Celsius above preindustrial levels, and the Fourth National Climate Assessment.

So human-induced warming reached approximately 1 degree Celsius, or 1.8 degrees Fahrenheit, a warmer world. And what has that brought us? Research indicates that this warming has changed the behavior and severity of extreme events.

For example, scientists found that global warming made the precipitation around 15 percent more intense for Hurricane Harvey that brought devastating flooding to Houston, and made it around three times more likely.

So, at the present rate, global warming would reach 1.5 degrees around 2040, and around 2 degrees around 2065. And every half a degree of global temperature increase has major consequences. For example, coral reefs have an immense variety of species and support fisheries that help feed many around the world. The Intergovernmental Panel on Climate Change Special Report assessed that coral reefs are projected to decline a further 70 to 90 percent at 1.5 degrees Celsius above preindustrial, and losses of nearly all coral reefs at 2 degrees Celsius above preindustrial levels.

To avoid surpassing 1.5 degrees Celsius, global carbon emissions would have to drop around 45 percent below 2010 levels by around 2030, and reach net-zero emissions by the mid-century. The special report asserts that to hold temperatures to 1.5 degrees would require "rapid and far reaching transitions in energy, land, urban, and infrastructure" at an "unprecedented scale" with "significant upscaling of investments in options." Given the scale of changes needed and the time to lay the framework, this is a make-or-break decade to make capital investments needed to reduce carbon dioxide levels, or the Paris Climate goals are unlikely to be achieved.

The Fourth National Climate Assessment was released in November in accordance with the legal mandate of the 1990 Global Change Research Act. And, increasingly, U.S. residents already recognize the consequences of climate change. Midwest forest products industry has experienced over the past 70 years 2- to 3-week shorter frozen ground season suitable for winter harvests. The Great Lakes ice cover decreased on average 71 percent from 1973 to 2010, with a recent rebound in the ice years of 2014 and 2015.

Meanwhile, during the 2012 and 2017 winters, in Lake Ontario and southern Lake Michigan the temperatures never dropped below 39 degrees Fahrenheit. And that's a critical threshold for seasonal mixing of the waters. Without winter or spring seasonal mixing, the chance is for increases for low oxygen conditions, which are toxic to aquatic species.

In another case, an extreme flooding event in Thailand caused a U.S.-based company to lose around half of its hard-drive shipments during the last quarter of 2011. Consumers may not have realized this, but this temporarily doubled global hard-drive prices and drove up the costs for Apple, HP, and Dell.

Climate change can exacerbate historical inequities. And I want to say that the projected costs in the labor is around $155 billion per year. And under a low-emissions scenario we could take a bite of nearly a half out of those damages. Extreme heat mortality could have damages towards the end of the century of over $140 billion per year. We could take a 48 percent bite.

Mr. TONKO. If I can ask you to wrap up, please.

Dr. EKWURZEL. And I just want to say overall coastal property losses, the losses are real, climate change is real. We need to step up solutions at the root cause, which States and cities are doing today.

Thank you very much.

[The prepared statement of Dr. Ekwurzel follows:]

Testimony of Dr. Brenda Ekwurzel, Director of Climate Science Union of Concerned Scientists, "Time for Action: Addressing the Environmental and Economic Effects of Climate Change," House Committee on Energy & Commerce, Subcommittee on Environment and Climate Change, February 6, 2019

Thank you, Chairman Tonko, Ranking Member Shimkus, and Members of the Subcommittee for providing me the opportunity to testify here today. I am Director of Climate Science at the Union of Concerned Scientists and served as a co-author of the Fourth National Climate Assessment released last November. Before I share with you advances in our understanding of climate from the latest climate assessments, I will turn to a recent example of the high cost of climate change.

During the recent outbreak of extreme cold weather that gripped large parts of the nation, a University of Iowa student and a University of Vermont student were counted among at least 21 people who perished from consequences likely related to the dangerous cold temperatures and wind chill.[1,2] Although it may seem counterintuitive, recent studies indicate that the warming climate can cause unusually cold temperatures at mid-latitudes by disrupting the normal winter season polar vortex in the stratosphere.[3] A good analogy to this disruption is a weak seal on a freezer door that periodically allows frigid air to flood into the room while warmer air rushes into the freezer. At the end of January, a cold blast spilled out of the Polar Regions and into the Midwest and eastern US – breaking wind chill records. Yet Alaska experienced a warmer-than-normal season, where above-freezing temperatures and rainfall forced the cancellation of mid-distance dog sled races that mushers use to qualify for long-distance races such as the Iditarod.[4] Evidence is growing that warmer-than-normal periods in the Arctic are associated with a greater chance for extreme winter weather in the eastern US.[5] This deadly cold snap is just the most recent example of the changing nature of extreme events that scientists are studying. One goal is to provide earlier warning so local officials have more time to take precautionary measures to improve safety.

Climate assessments provide the public and policy makers the most advanced warnings through summary and evaluation of the most recent research. I will briefly share some findings with you today from the Intergovernmental Panel on Climate Change Special Report on Global Warming of 1.5 degrees Celsius and the Fourth National Climate Assessment.

---

[1] Iowa City Press-Citizen Feb 2, 2019 by Aimee Breaux https://bit.ly/2Gk6feW.

[2] BBC News, Feb 1, 2019 https://bbc.in/2Sn104s.

[3] Polar vortex and Sudden Stratospheric Warming defined: https://www.aer.com/glossary/#P.

[4] KTUU Jan 29, 2019 by Jill Burke https://bit.ly/2Bnkdt7.

[5] Cohen, Pfeifer, and Francis, 2018, Nature Communications; https://www.nature.com/articles/s41467-018-02992-9.

Intergovernmental Panel on Climate Change Special Report Global Warming of 1.5°C

Nearly all countries have agreed to the goals of the Paris Climate Agreement, "Holding the increase in the global average temperature to well below 2°C above pre-industrial levels and pursuing efforts to limit the temperature increase to 1.5°C above preindustrial levels."[6] The Intergovernmental Panel on Climate Change was invited to issue a special report on the climate impacts of 1.5 degrees Celsius and different pathways that could limit the temperature increase. Human-induced warming reached approximately 1°C (1.8°F) above pre-industrial levels in around 2017.[7] What has this 1°C warmer world already brought us? Research indicates that this warming has changed the behavior and severity of extreme events. For example, scientists found that global warming made the precipitation around 15% (8-19%) more intense for Hurricane Harvey that brought devastating flooding to Houston, and around three (1.5-5) times more likely.[8]

At the present rate, global temperatures would reach 1.5°C around 2040 and 2°C around 2065.[9] Every half-degree of global temperature increase can have major consequences. Coral reefs have an immense variety of species and support fisheries that help feed many around the world. The IPCC special report assessed that coral reefs are projected to decline by a further 70-90% at 1.5 degrees Celsius with larger losses of nearly all coral reefs at 2 degrees Celsius above pre-industrial levels.[10]

---

[6] UNFCCC https://unfccc.int/process-and-meetings/the-paris-agreement/the-paris-agreement.

[7] IPCC SR15 SPM: "Human activities are estimated to have caused approximately 1.0°C of global warming5 above pre-industrial levels, with a likely range of 0.8°C to 1.2°C. https://www.ipcc.ch/sr15/.

[8] van Oldenborgh, G.J., van der Wiel, K., Sebastian, A., Singh, R., Arrighi, J., Otto, F.E.L., Haustein, K., Li, S., Vecchi, G. and Cullen, H. (2017) Attribution of extreme rainfall from Hurricane Harvey, August 2017. *Environmental Research Letters,* 12:124009. doi: 10.1088/1748-9326/aa9ef2.

[9] IPCC SR15 2018: "Global warming is likely to reach 1.5°C between 2030 and 2052 if it continues to increase at the current rate. (high confidence) {1.2, Figure SPM.1}" https://www.ipcc.ch/sr15/.

[10] IPCC SR15 2018.

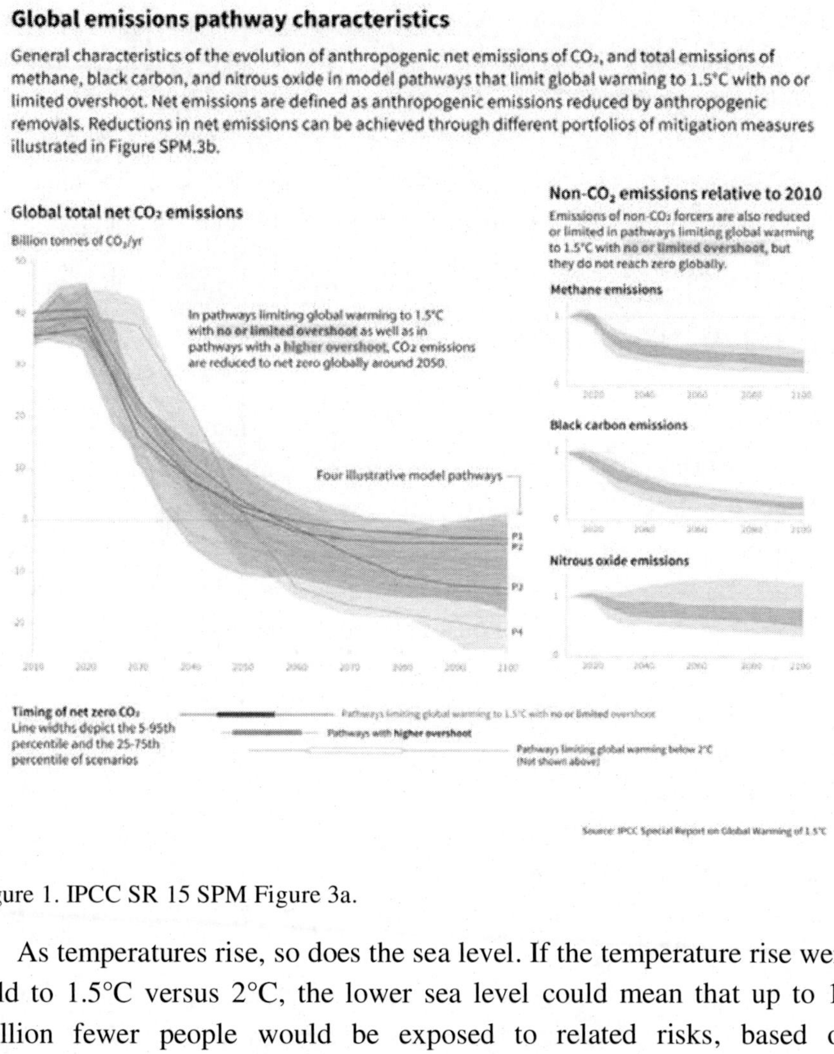

Figure 1. IPCC SR 15 SPM Figure 3a.

As temperatures rise, so does the sea level. If the temperature rise were held to 1.5°C versus 2°C, the lower sea level could mean that up to 10 million fewer people would be exposed to related risks, based on population in the year 2010 and assuming no adaptation.[11]

To avoid surpassing 1.5°C, global carbon dioxide emissions would have to drop around 45% below 2010 levels by 2030 and reach net zero emissions by around 2050 (Figure 1). This would also include deep reductions in methane and black carbon (or soot) as well as nitrous oxide (such as from agriculture). The special report asserts that to hold increasing

[11] IPCC SR15 2018.

temperatures to 1.5°C would require "rapid and far reaching transitions in energy, land, urban and infrastructure" at "unprecedented scale" with "significant upscaling of investments in options."[12] Given the scale of the changes needed and the time to lay the framework, this is the "make or break" decade to make the capital investments needed to reduce the carbon dioxide levels or the Paris Climate goals cannot be attained.

The Fourth National Climate Assessment

The Fourth National Climate Assessment, Volume II, was released in November 2018 in accordance with the legal mandate of the 1990 Global Change Research Act.[13] Major themes emerge from the findings. Increasingly U.S. residents recognize the consequences of climate change in their daily lives, their communities and livelihoods.[14] Social and economic inequities can be exacerbated with climate change through the increased exposure and sensitivity to extreme weather and other climate-related events or changes.[15] Actions taken to address the causes and impacts can affect disadvantaged populations unless equity dimensions are factored in. Cities and states are already deploying a mix of solutions to reduce emissions[16] and the iterative process of adaptation has progressed[17] since the third national assessment.

Increasingly U.S. residents already recognize the consequences of climate change. Midwest forest products industry has experienced over the past 70 years a 2-3 week shorter frozen ground season suitable for winter harvests.[18] Great Lakes ice cover decreased on average 71% from 1973 to 2010, with a recent rebound to higher ice years during 2014 and 2015.[19] Meanwhile, during the 2012 and 2017 winters in Lake Ontario and

[12] IPCC SR15 2018.
[13] NCA4 2018 https://nca2018.globalchange.gov/chapter/appendix-1/.
[14] NCA4 2018 https://nca2018.globalchange.gov/chapter/1/.
[15] Ibid.
[16] NCA4 2018 https://nca2018.globalchange.gov/chapter/29/.
[17] NCA4 2018 https://nca2018.globalchange.gov/chapter/28/.
[18] NCA4 2018 https://nca2018.globalchange.gov/chapter/21/.
[19] Ibid.

southern Lake Michigan, the temperatures never dropped below 39 °F, the critical temperature threshold for seasonal mixing.[20] Without the winter or spring seasonal mixing, the chances increase for low oxygen conditions that prove toxic for aquatic species. In another case, an extreme flooding event in Thailand caused a U.S.-based company to lose around half of its hard drive shipments during the last quarter of 2011. This temporarily doubled global hard drive prices and drove up costs for Apple, HP, and Dell.[21]

Climate Change can exacerbate historical inequities unless decision makers identify solutions that consider these factors. Low-income communities, children, older adults, and people of color are often at greater risk. Low-income communities are often located in areas that may be more prone to flooding (e.g., the Lower Ninth Ward during Hurricane Katrina). Extra attention must be paid to ensure that the vulnerabilities of frontline communities are identified and addressed. The health impacts of Climate Change can be mitigated to some degree. Adults can protect children from extreme heat through hydration and exertion monitoring. Arizona and California have a spike in cases of a fungus that can lead to Valley Fever during dry winter and spring conditions conducive to its growth. Over 40% end up hospitalized and two-thirds can take weeks to months before patients can resume normal activities.[22] Other health consequences of dust related to droughts and links with asthma, allergens and respiratory issues were more negative when affecting household property and finances.[23]

One of the advances of the Fourth National Climate Assessment was the integrated assessment of damages to the United States from particular levels of projected global emissions. Damages at the highest emissions trajectories were compared with “avoided damages” at lower global emissions trajectories (Figure 2). I will share the findings for the top three sectors. Projected annual economic damages (in 2015 dollars) for the labor sector in 2090 under a higher global emissions scenario (RCP 8.5) was on average $155 ($87-$221) billion per year. Under a lower global emissions

[20] Ibid.
[21] NCA4 2018 https://nca2018.globalchange.gov/chapter/16/.
[22] NCA4 2018 https://nca2018.globalchange.gov/chapter/14/.
[23] Ibid.

scenario (RCP 4.5) around 48% (40%-60%) average damages can be avoided. Projected annual economic damages for extreme heat mortality in 2090 under a higher global emissions scenario was on average $141 ($82-$201) billion per year. Under a lower global emissions scenario (RCP 4.5) around 58% (44%-69%) average damages can be avoided. Projected annual economic damages for coastal property in 2090 under a higher global emissions scenario was on average $118 billion per year. Under a lower global emissions scenario (RCP 4.5) around 22% average damages can be avoided.

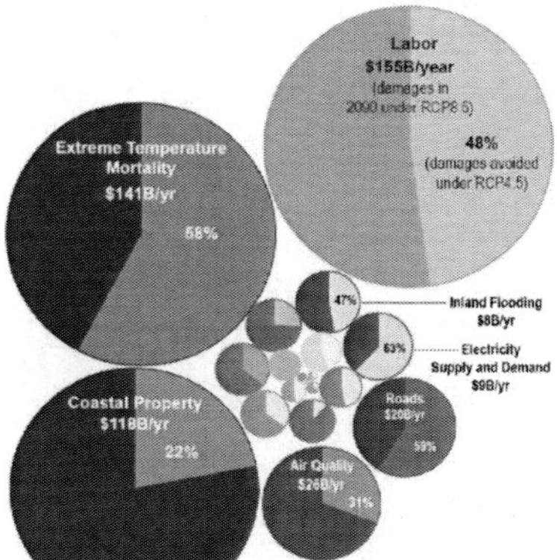

Annual Economic Damages in 2090		
Sector	Annual damages under RCP8.5	Damages avoided under RCP4.5
Labor	$155B	48%
Extreme Temperature Mortality◊	$141B	58%
Coastal Property◊	$118B	22%
Air Quality	$26B	31%
Roads◊	$20B	59%
Electricity Supply and Demand	$9B	63%
Inland Flooding	$8B	47%
Urban Drainage	$6B	26%
Rail◊	$6B	36%
Water Quality	$5B	35%
Coral Reefs	$4B	12%
West Nile Virus	$3B	47%
Freshwater Fish	$3B	44%
Winter Recreation	$2B	107%
Bridges	$1B	48%
Munic. and Industrial Water Supply	$316M	33%
Harmful Algal Blooms	$199M	45%
Alaska Infrastructure◊	$174M	53%
Shellfish*	$23M	57%
Agriculture*	$12M	11%
Aeroallergens*	$1M	57%
Wildfire	-$106M	-134%

Figure 2. NCA4 2018 Figure 29.2.

Climate Change is real and we are feeling its effects now. NOAA and NASA report that 2018 is among the top hottest years on record. According to UN population statistics, more than 60% of people in 2018 have experienced above average global temperatures every year of their lives while teenagers have experienced mostly record-breaking years.

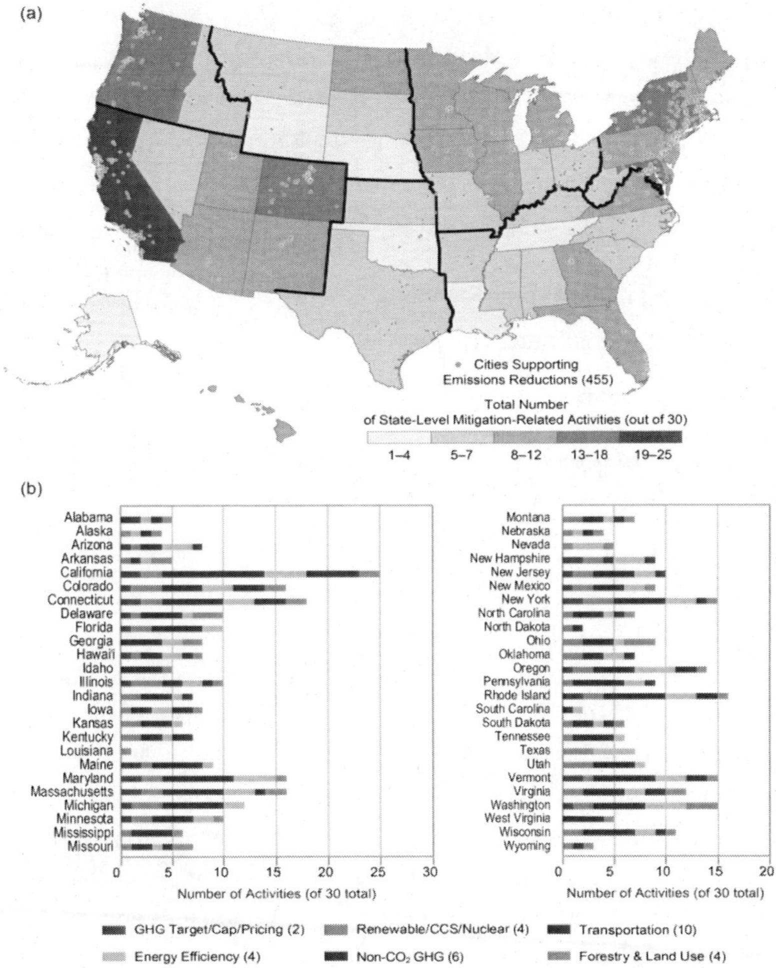

Figure 3. NCA4 Figure 29.1 (a) Map of mitigation-related activities at the state level and cities supporting emissions reductions; (b) type and number of activities by state. Several territories also have a variety of mitigation-related activities including American Samoa, the Federated States of Micronesia, Guam, Northern Mariana Islands, Puerto Rico, and the U.S. Virgin Islands. Sources (a) EPA and ERT, Inc. and (b) adapted from America's Climate Pledge 2017.

These two reports demonstrate that the lives of many Americans are at stake. The IPCC Special Report predicts the significant damage we will inflict on ourselves if we follow the high CO2 emissions trajectories while highlighting the real benefits, in terms of dollars and lives saved, of a lower emissions trajectory. Over 455 cities and all States are employing a mix of strategies to reduce the carbon dioxide emissions at the root cause of climate change (Figure 3).[24] We need to step up these and new efforts to protect and preserve a livable environment for ourselves, our children and grandchildren.

Thank you for this opportunity to speak with you today. I look forward to your questions.

Mr. TONKO. Thank you. And we now move to Mr. Rich Powell. You are recognized for 5 minutes, Mr. Powell.

STATEMENT OF RICHARD J. POWELL

Mr. POWELL. Good morning, Chairmen Tonko and Pallone, Republican leaders Shimkus and Walden, and other members of the committee. Thank you for the opportunity to appear today.

I am Rich Powell, Executive Director of ClearPath, a nonprofit that develops conservative policies that accelerate clean energy innovation. ClearPath supports flexible low-carbon technologies, nuclear, hydropower, carbon capture for both coal and gas, and energy storage.

Climate change is an urgent challenge that merits action at every level of the government and private sector. It is too important to be a partisan punching bag. Climate change deserves a pragmatic and technology-inclusive agenda to make the global clean-energy transition cheaper and faster. It is conservative to hedge for this risk.

Heavy industry is aggressively moving onto solutions to deal with climate issues. Southern Company is reducing their emissions in half by 2030, and will be low- to no-carbon by 2050. Shell also aims to cut emissions in half by 2050. Notably, senior executives from Southern, Shell, and just last week BP are linking their pay to hitting emissions targets. These examples illustrate that the Federal Government should enable private-sector solutions through market-oriented policies.

---

[24] NCA4 2018.

Crucially, we must also remember that climate change is a global problem. A molecule of CO2 emitted on the other side of the world has the same impact as one released here. Since 2000, coal power generation in China nearly quadrupled. Bloomberg reports that new Chinese coal capacity remains planned roughly equivalent to the entire U.S. coal fleet. Abroad, China is financing another 100 gigawatts of coal in at least 27 countries. The expected emissions growth from developing Asian countries by 2050 alone would offset a complete decarbonization of the U.S. economy.

More broadly, the share of global energy supplied by clean sources has not increased since 2005. Despite significant renewables growth, global emissions continue to rise. In other words, clean development is only just keeping up with economic development. Clean is not gaining ground. Clean tech available today is simply not up to the task of global decarbonization. It must represent a better, cheaper alternative so developing nations consistently choose it over higher-emitting options.

We have a choice: That the Chinese and their partners shut down their coal-fired power plants at the expense of economic growth, or develop, demo, and export U.S.-based emissions control technologies.

This technologies challenge is evident in the most ambitious plan yet from a major U.S. utility. Xcel Energy recently announced plans to reduce carbon emissions 80 percent by 2030 and 100 percent by 2050. Xcel noted they will require innovation to reach their 100 percent goal while remaining reliable and affordable for their customers. Growing their already impressive portfolio of renewables won't be enough.

A serious debate on climate solutions must include a dose of political and technical realism. Let's not rush toward any impracticably hasty, exclusively renewable strategy in the U.S. that will be both costly and unlikely to reduce global emissions. If supporters of a Green New Deal truly believe climate change is an existential threat, they should focus on policies that reduce global emissions as quickly and cheaply as possible.

So how do we change our trajectory? Well, we have done it before. There is no reason that clean technology needs to be more expensive or worse performing than higher-emitting technology.

Take America's shale gas revolution, rooted in decades of publicprivate research partnerships. This R&D, coupled with a $10 billion alternative production tax credit, yielded combined cycle turbines, diamond drill bits, horizontal drilling, and 3D imaging. Markets took up the technology, increasing gas from 19 to 32 percent of our power between 2005 and 2017, lowering emissions 28 percent.

The same ingenuity that produced the shale boom can make that gas fully clean. Near Houston, NET Power is successfully demonstrating a groundbreaking zero-emission natural gas power plant. More broadly, it is an immensely promising time for publicprivate partnerships in U.S. clean innovation. Some examples:

Form Energy is developing cheap, long-duration energy storage that may enable many more renewables. NuScale is licensing a small modular nuclear reactor, while Oklo and X–Energy partner with our national labs on microreactors.

The last Congress hasn't received the credit it is due for boosting low-carbon technologies. Your broadly bipartisan agenda enhanced critical incentives for carbon capture, renewables, and advanced nuclear, invested in clean R&D at record levels, and reformed regulations to accelerate the licensing of both advanced nuclear reactors and hydropower. One example: The 45Q tax incentive for carbon capture was supported by a vast bipartisan coalition, from environmentalists to labor to utilities to coal companies. Notably, seven national unions just collectively restated the need to include carbon capture and nuclear in any national climate policy.

Going forward, given the scale of the climate challenge, we need to greatly increase the pace and ambition of our efforts. Let's not shy away from smart investments in technology moonshots to deliver lost-cost, high-performing, clean technology. Let's create stronger incentives to commercialize cutting-edge companies and deploy their technologies globally, and remove regulatory barriers to rapidly scaling clean technology.

Bipartisan cooperation on climate change is essential under divided government, and attainable. In fact, it is the only chance our Nation will have to play a significant role in the global solution.

Thank you again for this opportunity, and I look forward to the discussion.

[The prepared statement of Mr. Powell follows:]

RICHARD J. POWELL, EXECUTIVE DIRECTOR, CLEARPATH, HOUSE ENERGY AND COMMERCE SUBCOMMITTEE ON ENVIRONMENT AND CLIMATE CHANGE, "TIME FOR ACTION: ADDRESSING THE ENVIRONMENTAL & ECONOMIC EFFECTS OF CLIMATE CHANGE"

Good morning Chairmen Tonko and Pallone, Republican Leaders Shimkus and Walden, and other members of the committee. Thank you for the opportunity to appear before you today and for holding this hearing.

My name is Rich Powell. I am the Executive Director of ClearPath, a non-profit that develops and advances conservative policies that accelerate clean energy innovation. My organization supports flexible low-carbon energy technologies - next-generation nuclear, hydropower, carbon capture on both coal and natural gas, and grid-scale energy storage.

Climate change is an urgent challenge that merits action at every level of government and the private sector. It is too important to be a partisan punching bag. Climate change deserves a pragmatic and technology-inclusive agenda to make the global clean energy transition cheaper and faster. It's conservative to hedge for this risk.

Heavy industry is aggressively moving onto solutions to deal with climate issues. Southern Company is reducing their emissions in half by 2030 and will be low to no carbon by 2050 - all while rapidly innovating clean tech. Shell also aims to cut its carbon emissions in half by 2050. Notably, senior executives from Southern, Shell, and just last week BP are among the growing list of big energy companies who are beginning to link future bonuses and other pay to their emission targets. These examples help illustrate a very clear principle: the federal government, where appropriate, should enable private-sector solutions through market-oriented policies.

Crucially, we must remember that climate change is a global problem. A molecule of CO2 emitted on the other side of the world has the same impact as one released here. Since 2000, coal power generation in China nearly quadrupled.[25] Bloomberg reports that over 250 gigawatts of new Chinese coal capacity remain planned, roughly the size of the entire U.S. coal fleet.[26] Abroad, China is financing another 100 gigawatts of coal in at least 27 countries.[27] The expected emissions growth from developing Asian countries alone would offset a complete decarbonization of the U.S. economy by mid-century.[28]

More broadly, the share of global energy supplied by clean sources has not increased since 2005. Despite significant global renewables deployment, emissions continue to rise. In other words, clean development is only just keeping up with economic development; clean is not gaining ground. Clean technology available today is simply not up to the task of global decarbonization. It must represent a better, cheaper alternative so developing nations consistently choose it over higher-emitting options. We have a choice - bet that the Chinese and their partners shut down their coal-fired power plants at the expense of economic growth; or develop, demonstrate, and export U.S.-based emissions control technologies.

This technology challenge is evident in the most ambitious plan yet from a major U.S. utility. Xcel Energy recently announced plans to reduce its carbon emissions 80% by 2030 and 100% by 2050. Xcel noted they will require innovative new technology to reach their 100% goal while remaining reliable and affordable for their customers. Their already-

---

25 https://www.shell.com/energy-and-innovation/the-energy-future/scenarios/shell-scenario-sky.html.

26 https://www.bloomberg.com/news/articles/2018-11-30/almost-half-of-coal-power-plants-seen-unprofitable-to-operate.

27 http://ieefa.org/wp-content/uploads/2019/01/China-at-a-Crossroads_January-2019.pdf.

28 https://www.eia.gov/outlooks/aeo/data/browser/#/?id=10-IEO2017&region=0-0&cases=Reference&start=2010&end=2050&f=A&linechart=Reference-d082317.3-10-IEO2017~~~~~~~Reference-d082317.17-10-IEO2017&map=&ctype=linechart&sourcekey=0.

impressive portfolio of existing renewable and other clean power isn't enough.

A serious debate on climate solutions must include a dose of political and technical realism. Let's not rush toward any impractically hasty, exclusively renewable strategy in the U.S. that will be both costly and unlikely to reduce global emissions. If supporters of a Green New Deal truly believe climate change is an existential threat, they should focus on policies that reduce global emissions as quickly and cheaply as possible.

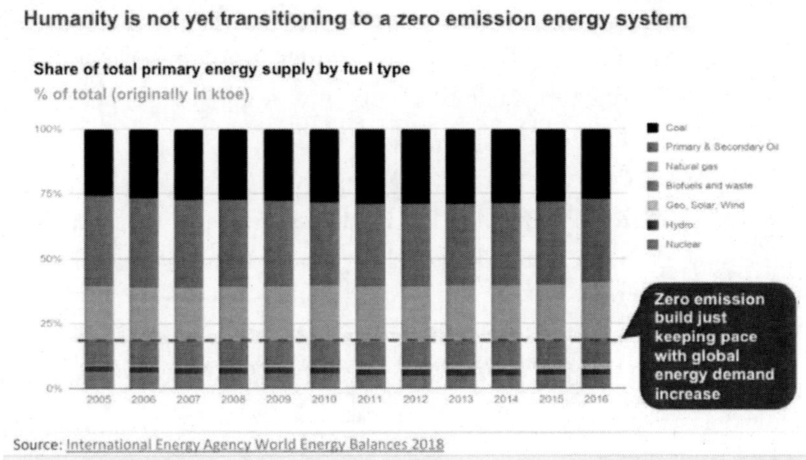

Figure 4.

So how do we change our trajectory? We've done it before. There's no reason that clean technology needs to be more expensive or worse performing than higher-emitting technology.

Take America's shale gas revolution, rooted in decades of public-private research partnerships.[29] This R&D, coupled with a $10 billion alternative production tax credit, yielded breakthroughs in combined cycle turbines, diamond drill bits, horizontal drilling, and 3D imaging.[30] This

---

[29] https://static.clearpath.org/2019/02/shale-gas-fracking-doc.pdf.

[30] http://americanenergyinnovation.org/wp-content/uploads/2013/03/Case-Unconventional-Gas.pdf.

market-driven phenomenon has increased natural gas from 19 to 32% of the grid[31] between 2005 and 2017, resulting in a 28% emissions decline.[32]

The same ingenuity that produced the shale boom can make that gas fully clean. A company called NET Power has created a groundbreaking zero-emission natural gas power plant. NET Power is successfully demonstrating near Houston and preparing to scale up. More broadly, it's an immensely promising time for U.S. clean innovation. Public-private efforts like Form Energy are developing cheap long-duration energy storage that may enable many more renewables. Intrepid entrepreneurs are innovating small modular nuclear reactors, such as NuScale, and micro-reactors such as Oklo and X-Energy, in partnership with our national labs. These efforts are representative of the aggressive public-private collaborations needed to dent this global problem.

The last Congress hasn't received the credit it is due for boosting low-carbon technologies. Your broadly bipartisan agenda enhanced critical incentives for carbon capture, renewables, and advanced nuclear; invested in Department of Energy R&D at record levels; and reformed regulations to accelerate the licensing of both advanced nuclear reactors and hydropower. The 45Q tax incentive for carbon capture and storage technology is a perfect example - it was supported by a vast bipartisan coalition from environmental organizations to organized labor to utilities to coal companies. Notably, seven national unions just collectively re-emphasized the importance of including carbon capture and nuclear in any national clean energy policy.

Going forward, given the scale of the climate challenge, we need to greatly increase the pace and ambition of our efforts. Let's not shy away from smart investments in "moonshot" goal programs that deliver low-cost, high-performing clean technology - from basic research all the way through demonstrations. Let's create stronger financing and incentives to commercialize cutting-edge companies and deploy those technologies globally. And let's enact deep regulatory reforms that remove barriers to rapidly scaling clean technology.

---

[31] https://www.eia.gov/survey/#eia-923.

[32] https://www.eia.gov/environment/emissions/carbon/.

Bipartisan cooperation on climate change is essential under divided government - and attainable. In fact, it is the only chance our nation will have if it is going to play a significant role in the global solution. Thank you again for this opportunity, and I look forward to the discussion.

Mr. TONKO. Thank you, Mr. Powell.

And next we will move to Mr. Rick Duke. You are recognized, Mr. Duke, for 5 minutes.

STATEMENT OF RICHARD D. DUKE

Mr. DUKE. Thank you, Chairman Tonko, Republic leader Shimkus, and members of the committee for inviting me to testify on the prospects for reducing greenhouse pollution through American leadership on technology and diplomacy. It is an honor to share with this committee my confidence that we can still contain the most costly and destabilizing climate impacts, but only if we choose to act to put our Nation on a path to net-zero greenhouse gas pollution by mid-century.

In short, rapid climate action is strategic for both our economy and our national security. And we urgently need strong Federal policy to make it all happen. This is a momentum game—the faster we act, the easier it gets. Early support for emerging green technologies gives American entrepreneurs the chance to cut costs as they scale up production and learn by doing. As these costs come down, bigger markets open up, including for exports to countries that raise their ambition in response. And this in turn allows further cost reductions in globalscale economies.

This virtuous cycle spurs the incredible progress we are seeing for climate solutions ranging from super-efficient lighting to renewables. And many of these originated in American labs and startups. To build on this momentum, we need to double down on cutting greenhouse gas pollution in the United States. And we know exactly what to do. It starts with quickly scaling up zero-carbon electricity. We have to broadly electrify vehicles, buildings, and much of industry, and we also have to cut non-CO_2 greenhouse gases.

Over time, solutions that remove carbon dioxide from the atmosphere will play an increasingly important role. This includes restoring farmlands and forests through increased economic productivity, while also storing carbon in healthier soils and vegetation. At the same time, we need to kick start promising emerging technologies to directly extract CO2 from the atmosphere and safely sequester it.

These carbon dioxide removal solutions will allow us to achieve net zero by balancing out certain emissions that we don't know how to eliminate currently, such as methane and nitrous oxide from agriculture.

Despite the imperative to get moving, though, some argue that other countries aren't doing much so we should hold off on cutting our emissions. But the facts are that our competitors are already moving. Every country other than the U.S. remains committed to the Paris Agreement. The EU and Canada both have carbon pricing in place that is strong.

Mexico is moving to 35 percent clean electricity by 2024. And China has over 80 strong technology deployment policies in place that are propelling up to nearly $130 billion in renewables investment in 2017 alone. That is triple the level in the U.S.

At the same time, China already accounts for well over half the electric vehicle sales, and two of the top three electric vehicle manufacturers in the world. Tesla is still in the number one slot, and GM is in the top ten.

All this investment is driving down low-carbon technology costs globally, including batteries and solar electricity, both of which have come down about 80 percent since 2010. It has never been easier to cut greenhouse gas pollution. And all 50 States can act now. In fact, at least 45 States have already installed utility-scale solar and wind at increasingly prices that are below conventional power. And we are making progress with carbon capture and storage, including the zero-carbon natural gas electricity pilot in Texas, and cleaner ethanol in the Midwest.

But, unfortunately, we are not moving fast enough. Last year our energy CO2 emissions were up over 3 percent after a decade of falling about 1.5 percent per year. And now Federal policy is creating headwinds. The last two budget proposals sought to cut energy R&D by as much as 70 percent. Thankfully, Congress strategically increased funding on a bipartisan basis.

On deployment, the current administration is seeking to gut the Clean Power Plan, weaken vehicle standards, thereby threatening to cost drivers billions at the pump in higher gasoline consumption, and undermining measures to cut energy waste and methane leaks from our oil and gas systems. Instead of rolling back standards, we need stronger Federal investment in policy, both new legislation and vigorous implementation of existing law, to propel all low-carbon solutions forward.

Many different policy packages could get the job done, but this ideally starts with at least doubling clean energy R&D, plus legislation that puts a price on pollution and equitably and productively uses resulting revenue. And we absolutely can and must do right by workers and others on the front lines of this transition, including those struggling with the decline of coal, and communities most impacted by pollution. Added all together, we could cut our emissions in half by 2035, on track to net zero by mid-century, while bolstering our technological and diplomatic leadership.

Thank you. Look forward to the discussion. [The prepared statement of Mr. Duke follows:]

Testimony of Richard D. Duke Principal, Gigaton Strategies LLC, US House of Representatives, Energy and Commerce Committee, Environment and Climate Change Subcommittee, Hearing entitled "Time for Action: Addressing the Environmental and Economic Effects of Climate Change," February 6, 2019

Thank you, Chairman Tonko, Ranking Member Shimkus, Vice-Chair Ruiz, and Members of the Subcommittee, for inviting me to testify on the prospects for mitigating greenhouse gas emissions through American leadership on technology, policy, and diplomacy. I am grateful for the opportunity to speak with you today about this crucial challenge and opportunity. I develop strategies to reduce greenhouse gas pollution through my consulting practice, Gigaton Strategies LLC, and it is an honor to share with this committee my *conditional* optimism that we can still contain the most costly and destabilizing climate change impact if we choose to put our nation on a path to *net zero* greenhouse gas pollution by midcentury.

The stakes are high. On current course we face steadily-worsening impacts like sea level rise punctuated by episodic climate-intensified crises including unprecedented floods, heat waves, and cold snaps. If we fail to act, the Fourth National Climate Assessment finalized last year projects annual losses in multiple sectors of the US economy reaching hundreds of billions of dollars by the end of the century, hitting low-income Americans hardest. Just last week the World Threat Assessment from the Director of National Intelligence underscored the role of climate change in fueling global "…competition for resources, economic distress, and social discontent…"[33] Climate change will impose the most acute suffering on the world's poorest and least stable countries, underscoring the moral and national security imperatives to contain this looming crisis.

Climate mitigation is a momentum game—the faster we act, the cheaper pollution-reducing technologies become. When we put in place policies and incentives that deploy emerging solutions like solar and wind power, they become cheaper through economies-of-scale and learning-by-doing across the entire value chain. This, in turn, opens up ever-larger markets for clean technologies, catalyzing a virtuous cycle between declining costs and expanding sales (Figure 5).[34] US climate technology leadership also encourages other countries to pick up the pace as they

---

[33] https://www.odni.gov/index.php/newsroom/congressional-testimonies/item/1947-statement-for-the-record-worldwide-threat-assessment-of-the-us-intelligence-community.

[34] Duke, Richard D. (2002). "Clean Energy Technology Buydowns: Economic Theory, Analytic Tools, and the Photovoltaics Case." Dissertation presented to faculty of Princeton University. Woodrow Wilson School of Public and International Affairs. http://rael.berkeley.edu/old_drupal/sites/default/files/very-old-site/PhD02-Duke.pdf.

benefit from the low-cost low-carbon solutions we commercialize and export.

Finally, moving fast ensures we invest from the start in smarter cities and energy systems rather than stranding investment in polluting infrastructure that we end up having to abandon early.

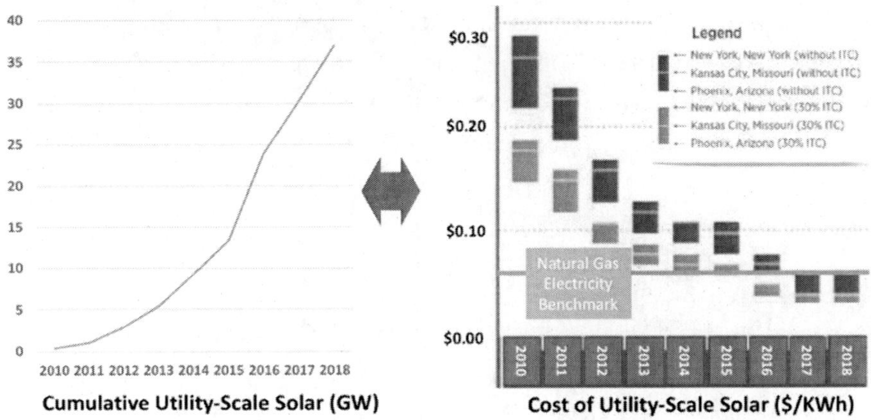

Figure 5. Virtuous cycle between expanding markets and cost reductions achieved through scale and learning-by-doing: the example of utility-scale solar in the U.S.[35]

We should prioritize rapidly reducing greenhouse gas pollution. This requires quickly scaling up zero-carbon electricity sources; broadly electrifying transportation, buildings, and industry; and, driving down major industrial and agricultural sources of methane and nitrous oxide. We know where the pollution comes from and the United States has pioneered key technologies to reduce it, but we need to move much faster to deploy these solutions.

Carbon dioxide (CO2) *sinks* will also play an increasingly important role in compensating for intractable greenhouse gas *sources* as we drive towards *net zero* emissions (Figure 6). There are two complementary types of CO2 sinks: *natural* and *technological*. Investing in land restoration to

---

[35] Solar cost trends from Q2/Q3 2018 NREL Solar Update accessible at https://www.energy.gov/eere/solar/quarterly-solar-industry-update, indicative natural gas electricity costs from bottom-end of range estimated by https://www.greentech media.com/articles/read/nevada-beat-arizona-record-low-solar-ppa-price#gs.dPxW6xwJ, and utility-scale solar volumes from https://www.seia.org/solar-industry-research-data.

create natural carbon sinks in forests, agricultural soils, and coastal ecosystems offers tremendous benefits, ranging from improved land productivity to enhanced resilience to climate impacts, but these opportunities are challenging to scale, involve reversibility risk (including from climate impacts like drought and fire), and are ultimately limited by total available land. Carbon dioxide removal technologies, such as direct air capture of CO2, are not similarly constrained, but they are much more expensive than near-term opportunities to reduce pollution. Thus, even as we prioritize immediate greenhouse gas reductions we must also invest in scaling land restoration and developing negative emissions technologies.

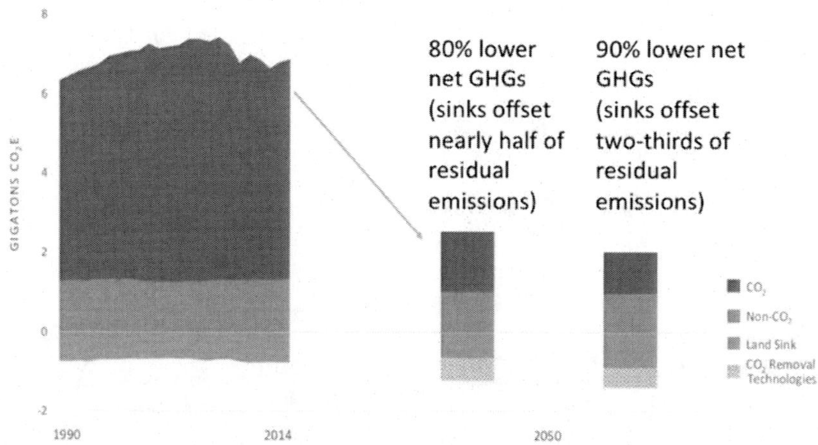

Figure 6. Land sinks and carbon dioxide removal technologies play a crucial role as the US drives toward net zero (United States Midcentury Strategy for Deep Decarbonization, 2016).[36]

We have the private sector dynamism and subnational policy commitment to lead this global economic transformation. But we urgently need federal policy progress to pick up the pace— both new legislation and vigorous implementation of existing legislative obligations and authorities.

[36] https://unfccc.int/files/focus/long-term.../mid_century_strategy_report-final_red.pdf.

Large-Scale Global Demand for Low-Carbon Technologies Is Driving Costs Down

There is a common, outdated, perception that cutting greenhouse gas pollution is expensive and other countries are therefore doing nothing about their emissions. In fact, every country other than the United States remains committed to the Paris Agreement and all of our major trading partners are taking action (Figure 7).

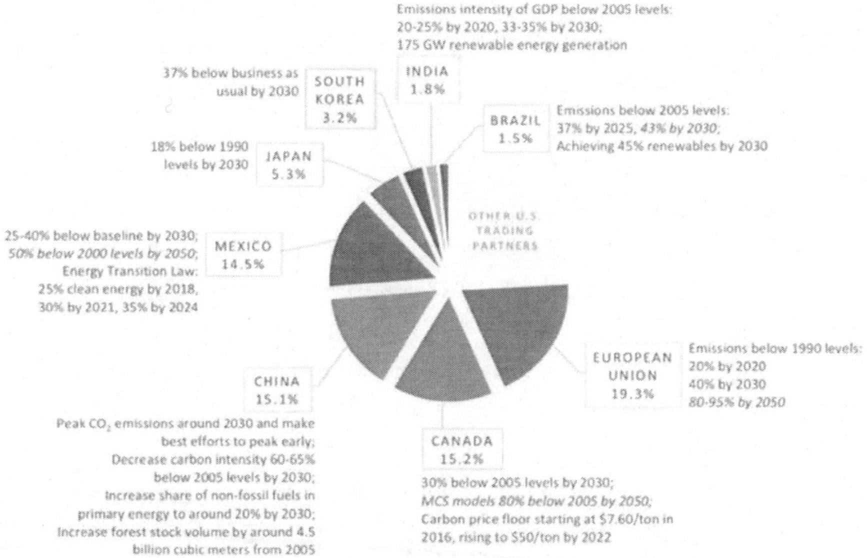

Figure 7. Our major trading partners are taking action (Segment size represents country's contribution to U.S. total trade volume).[37]

The European Union has far-ranging renewables and efficiency policies embedded in an emissions trading system currently imposing a carbon price of roughly $20-30 per million metric ton (MMT) of CO2. Canada has started to implement its Pan-Canadian Framework on Clean

[37] United States Midcentury Strategy for Deep Decarbonization, 2016 using Census Bureau data. Total trade equals the value of imports from country plus U.S. exports to country. Remainder of circle is comprised of other trading partners, the large majority of which have also developed NDCs.

Growth and Climate Change which includes a backstop carbon price rising to nearly $40 per MMT of CO2 by 2022 and a range of robust complementary innovation and deployment measures.[38] Under its 2015 Energy Transition Law, Mexico is competitively procuring renewable electricity to reach its goal of 35% clean electricity by 2024 and recently proposed new regulations to control oil and gas methane emissions.[39]

Since 2000, China has issued and implemented 83 major policies to scale up clean technologies, including sustained deployment incentives for renewable energy.[40] By 2017, China was investing $127B in renewable power and fuels, more than triple related US investments that year, and it already had more than twice as much total installed wind and solar capacity as in the U.S.[41] China also sold half a million electric vehicles (EVs) in 2017, half of the global total, and it is launching a "new energy vehicles" tradable credit mandate this year expected to triple EV sales to 1.5 million by 2020.[42] Two of the top three global EV manufacturers are Chinese firms, though Tesla retains the number one slot and General Motors is in the top ten.[43] China is also tracking to deliver on its 2014 pledge, jointly announced with the United States, to more than double the share of its total energy from carbon-free sources and peak its emissions by around 2030.

Similarly, India is installing more new wind and solar energy than new fossil-based generation, and independent analysts estimate that this surge of non-hydro renewables will deliver 13 percent of total electricity by 2023.[44] According to Bloomberg New Energy Finance (BNEF), new wind

---

[38] http://publications.gc.ca/collections/collection_2017/eccc/En4-294-2016-eng.pdf.

[39] http://www.lse.ac.uk/GranthamInstitute/law/energy-transition-law/ and https://www.edf.org/media/mexico-unveils-practical-oil-and-gas-regs-cut-climate-damaging-methane.

[40] Kelly Sims Gallagher and Xiaowei Xuan (2019); Titans of the Climate: Explaining Policy Process in the United States and China.

[41] Renewable energy excludes large hydro in this report. http://www.ren21.net/status-of-renewables/global-status-report/.

[42] http://www.ren21.net/status-of-renewables/global-status-report/ and https://www.bloomberg.com/news/articles/2018-11-14/china-is-about-to-shake-up-the-world-of-electric-cars-quicktake.

[43] https://insideevs.com/global-sales-in-december-full-year-2018-2-million-plug-in-cars-sold/amp/.

[44] https://www.greentechmedia.com/articles/read/woodmac-expects-india-to-miss-2022-renewables-target#gs.qEite3Nt.

and solar are both about $0.04 per kWh in India—much cheaper than new coal at $0.07/kwh and new combined cycle gas at 0.09/kwh.[45]

All of this investment is driving down the cost of low-carbon technologies through economies of scale and learning-by-doing. For example, global average costs for battery and solar electricity costs are both down about 80 percent since 2010.[46] This translates into affordable mitigation opportunities in the United States as well.

It Has Never Been Easier to Cut Greenhouse Gas Pollution, and All 50 U.S. States Have Opportunities to Lead

In addition to sustained investment in world-leading research, the United States benefits from a rich and diverse resource base, a dynamic private sector, and innovative subnational governments. Building on these strengths and global technology cost reduction trends, cutting emissions in the United States has never been easier, and every state has opportunities to lead the transition to a clean economy.

Figure 8 highlights dramatic reductions in the average cost of solar and wind electricity in the United States such that both now readily compete with new coal and natural gas generation and increasingly compete with existing coal power in certain markets.

Consistent with these national trends, renewables are becoming increasingly competitive in all 50 states (Figure 9), with a growing number of states competitively procuring solar at prices in the range of $0.02/KWh, down from over $0.15/KWh a decade ago and well below conventional power.[47] At the same time, over 40 states have installed utility-scale wind power, with contracted prices falling from $0.07/KWh in 2009 to $.02/KWh in 2017.[48] Wind is becoming competitive in ever-broader geographies with advancements like taller turbines and offshore

---

[45] https://bnef.turtl.co/story/neo2018?teaser=true.
[46] https://bnef.turtl.co/story/neo2018?teaser=true.
[47] https://www.greentechmedia.com/articles/read/nevada-beat-arizona-record-low-solar-ppa-price #gs.wvWKRR3y.
[48] https://www.energy.gov/eere/wind/downloads/2017-wind-technologies-market-report.

installations opening up less windy regions like the Atlantic Seaboard and Southeast. Nextera, the largest utility in the country with renewable installations spanning dozens of states, predicts that by the 2020s, unsubsidized solar and wind backed with some storage will be cheaper than conventional power, without the emissions or fuel price risk.[49]

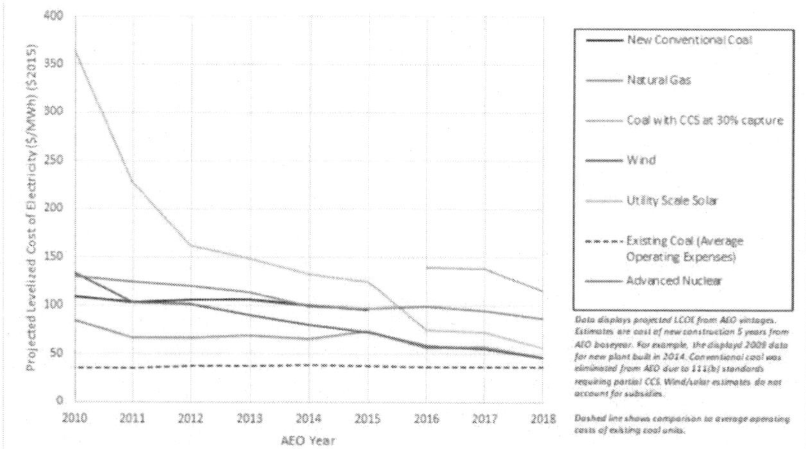

Figure 8. Electricity from low-carbon technology is now competitive with new conventional coal and natural gas derived power in the United States (derived from EIA).

Multiple states and industries are also pushing ahead with advanced carbon capture and storage solutions. This key technology is not only useful in the power sector but also essential for cutting industrial pollution and creating options for scaled carbon dioxide removal in coming decades. Netpower is piloting carbon capture and storage technology in Texas with the goal of producing carbon-free power from fossil fuels at about the same cost as conventional power.[50] At the same time, ethanol producers in the

49 https://www.statista.com/statistics/237773/the-largest-electric-utilities-in-the-us-based-on-market-value, http://www.nexteraenergy.com/company/work.html and http://ieefa.org/nextera-solar-wind-and-storage-will-be-massively-disruptive/.

50 https://qz.com/1292891/net-powers-has-successfully-fired-up-its-zero-emissions-fossil-fuel-power-plant/.

Midwest are moving to cut their lifecycle carbon emissions in half through carbon capture and storage on their refineries.[51]

Our farmers, including in the Midwest, are leading the world in advanced technologies that boost productivity and drive down greenhouse gas emissions through solutions like optimal fertilizer application and converting methane from waste to energy. And our forests in the Southeast and other forested regions already absorb over 10 percent of our emissions and could cost-effectively offset over half of our residual emissions by midcentury.

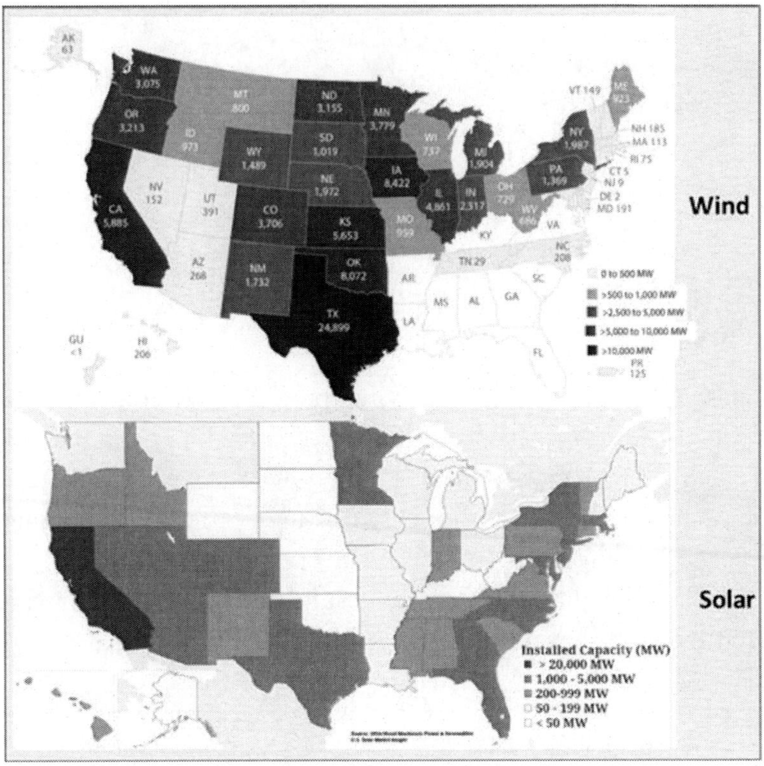

Figure 9. Wind and solar can compete in every state (cumulative installed MW).[52]

[51] https://www.adm.com/news/news-releases/adm-begins-operations-for-second-carbon-capture-and-storage-project-1.

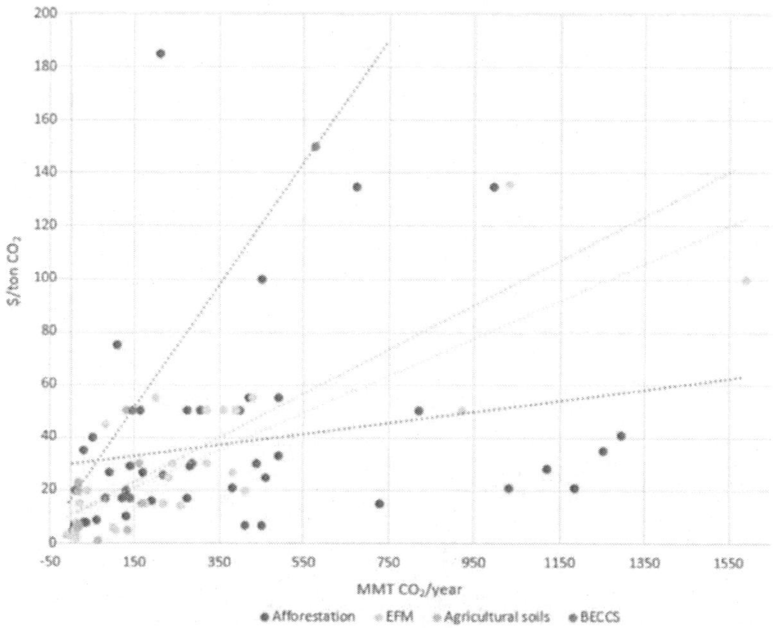

Figure 10. Potential US land sinks by midcentury through afforestation, enhanced forest management (EFM), soil management, and bioenergy carbon capture and storage (BECCS).[53]

States and cities are also driving progress, including through the bipartisan US Climate Alliance that recently added Virginia, New Jersey, Illinois, and New Mexico to include a total of 18 states and Puerto Rico.[54] A dozen states accounting for roughly one-third or new vehicle sales have opted to adopt California's greenhouse gas performance standards for light duty vehicles and plan to proceed despite the current administration's proposed federal rollbacks.[55] At the same time, the Regional Greenhouse

---

52 https://www.awea.org/resources/publications-and-reports/market-reports/2018-u-s-wind-industry-market-reports/4q2018_public and https://www.seia.org/solar-industry-research-data.

53 Based on 13 studies (synthesized in Van Winkle et al., 2017; Murray et al., 2005; Chambers et al., 2016) compiled by Emily McGlynn, https://www.rmi.org/hot-property-how-to-manage-valuable-us-landscapes-for-carbon-sequestration/.

54 https://www.usclimatealliance.org/governors-1/ and https://www.usclimatealliance.org/advanced-transportation/.

55 https://eelp.law.harvard.edu/2018/11/beyond-the-waiver-epas-plan-to-end-state-ghg-vehicle-standards/.

Gas Initiative is expanding beyond the power sector to tackle transportation and many states are propelling zero emission vehicles through a range of standards and incentives.[56] This subnational policy gives firms like Tesla and General Motors a path to continue scaling up production, hopefully ensuring that the majority of EVs on our roads continue to be made in America.

The United States Is Moving Too Slowly and Risks Losing Its Clean Technology Edge

Despite this global technology momentum and subnational climate policy, we are not moving fast enough. Energy CO2 emissions in the US, which accounts for over four-fifths of greenhouse gases, were up 3.4% in 2018 after falling 1.6% per year on average from 2007 to 2016, and only 0.8% in 2017.[57] As shown in Figure 11, the power sector has accounted for most of this downward emissions trend since 2007.

It is worth a closer look at the power sector to unpack three factors that propelled these emission reductions and to consider whether we are on pace even in this best-case sector (Figure 12). First, energy efficiency has held total electric load growth roughly level even as the economy expanded. Federal investment in energy saving technology paired with appliance and equipment standards have combined powerfully with subnational actions, including building codes and demand-side management programs. As one example, DOE has invested about $0.4B to develop American leadership on solid-state lighting such as LEDs, yielding nearly $5B in energy savings already and potentially $50B annually by 2035 as deployment scales.[58] Going forward, the building sector can also become a crucial source of demand flexibility, along with electric vehicles,

---

[56] https://www.utilitydive.com/news/regional-initiative-to-reduce-transportation-emissions-would-mirror-rggi/544738/.

[57] https://www.epa.gov/ghgemissions/inventory-us-greenhouse-gas-emissions-and-sinks and https://rhg.com/research/preliminary-us-emissions-estimates-for-2018/.

[58] https://www.energy.gov/eere/ssl/downloads/led-efficacy-what-america-stands-gain.

to ease large-scale integration of large intermittent renewables into our electricity systems.

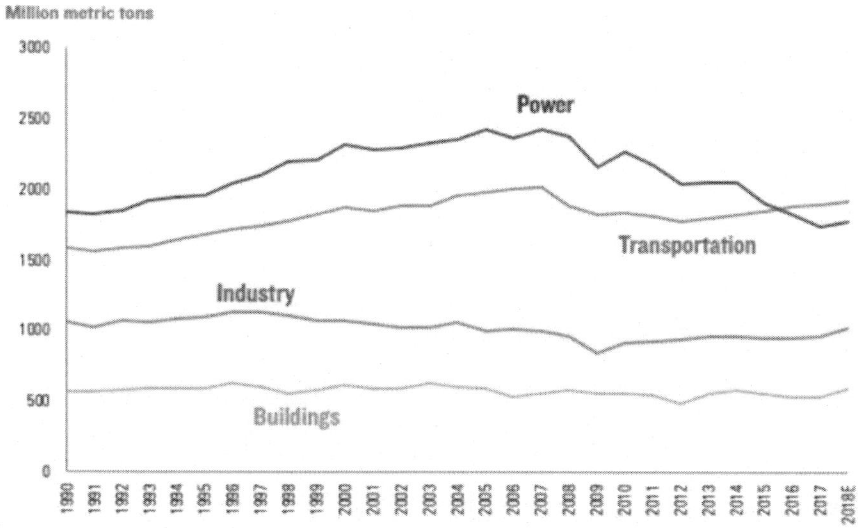

Figure 11. Transportation has overtaken power as the largest source of energy CO2 (RHG).

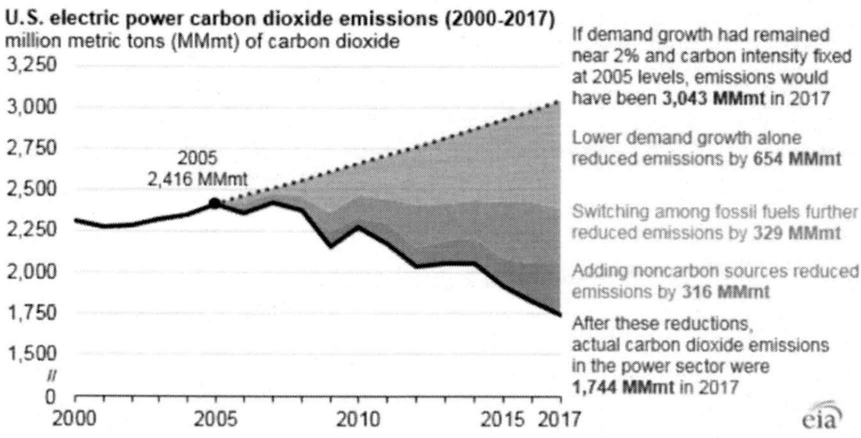

Figure 12. Underlying drivers of recent power sector emissions reductions.[59]

[59] https://www.eia.gov/todayinenergy/detail.php?id=37392.

Second, coal-to-gas shifts were a driver for power sector emissions reductions over the past decade; however, this dynamic is unlikely to play a major role going forward. Federal investment in hydraulic fracturing helped industry to increase natural gas supply during this period, driving down natural gas prices which, in turn, induced switching from coal to natural gas electricity. Going forward, EIA expects natural gas prices to trend higher, causing capacity utilization rates for coal-fired generators to revert to historical levels of about 70% from the low of 55% reached in 2018.[60] Moreover, improved monitoring indicates that methane emissions from the oil and gas supply-chain may be 60% higher than current official EPA estimates, principally due to leakage during abnormal events.[61] Absent standards that require more leak detection and repair, the near-term warming impact from methane leaks could rival the impact of total U.S. natural gas combustion.

Thus, assuming efficiency continues to largely offset load growth, rapidly expanding renewables will emerge as the dominant driver of durable reductions in power sector emissions. In fact, renewables have been rapidly gaining market share, and EIA projects this will continue through at least 2020 (Figure 13).

Even in the power sector, however, we are not moving fast enough. The MCS indicates that we would need to roughly double the pace of renewables installations starting in the 2020s to optimally cut net greenhouse gas emissions by 80 percent below 2005 levels by 2050. The pace would of course need to quicken further to achieve net zero by that year. Unfortunately, after federal renewable energy tax credits phase-out in the early 2020s, industry analysts project a serious downturn, with EIA estimating that new wind installations may even falter entirely in 2022.[62] Meanwhile, emissions are trending higher in the crucial transport sector and at best level in other sectors, even before federal policy reversals impose new headwinds.

---

[60] https://www.eia.gov/outlooks/aeo/.

[61] http://science.sciencemag.org/content/361/6398/186.

[62] https://www.eia.gov/outlooks/aeo/ and https://www.greentechmedia.com/articles/read/5-factors-that-will-cushion-the-ptc-phase-us-wind#gs.2sDg96ly.

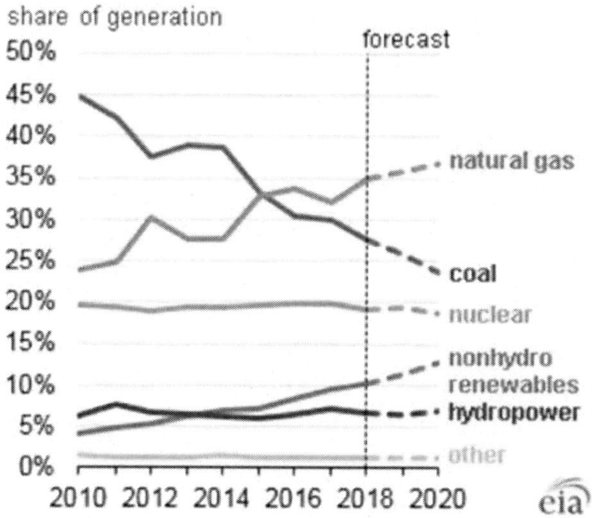

Figure 13. Wind and solar are rapidly gaining market share (EIA).

The United States Urgently Needs Federal Policy Progress to Retain Its Global Leadership on Climate Technology and Diplomacy

Even as the falling cost of clean technologies further bolsters the economic case for action, federal climate policy is shifting into reverse. On innovation, the current administration has repeatedly sought to slash investment in advanced clean energy technologies, though Congress has thus far maintained and even increased related investment. Most recently, for FY19 Congress finalized modest increases in DOE research and development investments despite Administration requests for cuts as severe as 70% for key programs.[63]

On deployment, the current administration is aggressively seeking to roll back pollution control standards. Jeopardizing power sector progress, it is seeking to undermine common sense standards seeking to level the playing field between conventional fossil generators and clean power

---

[63] https://www.aaas.org/page/fy-2019-rd-appropriations-dashboard.

alternatives, including renewables, nuclear, and advanced fossil with carbon capture and storage. These efforts include gutting the Clean Power Plan and repeated efforts to subsidize coal generation under the guise of reliability. Moreover, while states are leading the way in clean electricity policy, the Federal Energy Regulatory Commission is actively discouraging such policies by raising the costs to consumers with "minimum offer" requirements on sales from clean energy sources.[64]

The current administration is also moving to undermine vehicle standards that would save drivers billions at the pump, empower US automakers to compete with China for the EV future and keep the US roughly on track to deep decarbonization by midcentury.[65] The administration is similarly seeking to weaken common sense standards aimed at reducing leaks of methane from oil and gas production and distribution. It is also seeking to restrict the process for setting future appliance standards that benefit consumers.[66] This is not to mention persistent attempts to erode other critical public health standards such as the Mercury and Air Toxics Rule.

The United States remains uniquely well-positioned to lead the global effort to contain climate change, but we must take decisive steps to give investors the policy clarity needed to scale up solutions. Instead of rolling back standards, we need to ratchet up policy ambition to get the US back in a leadership role, motivating other countries and exporting the climate solutions that allow them to cut emissions faster.

This challenge is too important to leave any tool unused. We need broad, performance-based policies that propel investment to electrify end uses with clean power while catalyzing faster progress in precision

---

[64] https://www.greentechmedia.com/articles/read/could-ferc-order-put-state-clean-energy-policies-in-danger#gs.8iFTRGmj.

[65] From the US Midcentury Strategy (2016), "…if the current expected trajectory of emissions intensity improvements due to fuel economy and GHG emissions standards is sustained through 2050, fleet-wide emissions intensity would decline 76 percent between 2015 and 2050. In the MCS Benchmark scenario, emissions intensity declines 86 percent over the same period. Thus, with only a slight acceleration compared to current trends, fuel economy and GHG emissions standards have the potential to achieve carbon pollution reductions consistent with a deeply decarbonized energy system."

[66] https://www.greenbuildingadvisor.com/article/federal-agencies-begin-efficiency-policy-roll-backs.

agriculture, advanced manufacturing, and improved land management. We also need to invest in coal mining communities and others on the frontlines of this transition to ensure all Americans benefit, including through programs like the Power Plus initiative.[67] This will unleash a wave of reinvestment in American industry, agriculture and forestry that will ensure sustained competitive advantage.

Add it all together and we could cut our net emissions in half by 2035, on track to net zero by midcentury while propelling economic growth and ensuring America retains its leadership in related technologies and international negotiations.

Mr. TONKO. Thank you very much, Mr. Duke.

And now we will move to Reverend Leo Woodberry. Reverend, you are recognized for 5 minutes.

STATEMENT OF REVEREND LEO WOODBERRY

Reverend WOODBERRY. Thank you. Thank you, Chairman Tonko, and thank you distinguished members of the committee.

I have been doing this work now for over 25 years dealing with issues of climate and environmental justice. I could begin by talking about being too big to fail. But if we talk about that, then we can also talk about how we should not have moved away from kerosene to electric lights, or how we should have protected the carriage and buggy whip industry rather than developing the auto industry. Or we could have said, et's keep the typewriter industry going and never develop a computer industry.

So those are topics I can talk about. But what I would like to talk about is what we found last year when we conducted the Justice First Tour and went through 12 southeastern States and 25 cities and talked to people on the front line, people who have been suffering the impacts of carbon emissions, pollution, and the impacts of climate change.

---

[67] See, for example, http://www.powerplusplan.org/.

So I am talking about people like the 90-year-old woman in Sellers, South Carolina, in Marion County who now has to elevate her home 7 feet in the air.

I am talking about people who labored in our fields, cleaned our homes, and worked for employers who never paid into their Social Security and have to live off SSI checks of $600 and $800 a month.

These are the people who are being impacted. We don't have to wait 12 years for a switch to be flipped. Americans are suffering the impacts of climate change right now. People being displaced, communities are being destroyed. And we come here issuing the clear clarion call of hope. We need policy change. We need to desperately put our people to work.

We can, like in the town of Sellers, South Caroline, they said that the flooding impacts were worse because of large-scale logging, losing our natural defenses against flooding. Because the ditches had not been cleaned out in 25 years in this rural community.

We can put our people to work elevating homes, cleaning out ditches, building bioswales to minimize flooding. We can pass legislation that will put in place community-based climate solutions. It is time to move beyond the false narrative that equates big utilities with renewable energy.

Let's look at the justification. Utilities said, "We could not exist in a competitive environment because we have to build such large infrastructure that we might not get a return on our investment." Solar and wind can exist in a competitive environment. We don't have to look just towards macro solutions. If we can put timers and do energy efficiency in 10 million homes and reduce energy generation by as little as 200 kilowatt hours a year, we will have made a significant difference. But in order to do this we have to be able to look towards people who desperately need work.

We have counties, like Marion County, like Dillon County, like Darlington County, like counties all across this country, rural communities where people have to drive 25, 30, 40 miles each way every day because there are no engines of economic development in their community.

I came here today to talk about the people along the Black Belt, the people of Flint, Michigan, the people along the I–95 corridor of shame, the least among us, those who were forgotten about, who we turned our gaze away from while the same polluting facilities were allowed to be sited in their communities that have led to climate change, and the possibility of humanity no longer having civilization as we know it. We can debate forever whether or not climate change is real. But the problem is here. The problem is now. And we need to build a wall of protection around the citizens of this country, a wall of mitigation, a wall of adaptation, and a wall of resilience.

Because the science is clear, whether we are looking at the Intergovernmental Panel on Climate Change or we are looking at our own National Climate Assessment, the storms are going to get worse. The hurricanes are going to become more intense. And we have to keep our forests standing in the ground because they are the greatest carbon sinks on this planet. And we don't have enough time to see whether or not some technologies might work.

Mr. TONKO. Reverend, if you could wrap up.

Reverend WOODBERRY. And so I just want to close by saying this:

The time for action is now. And if we don't take action today, then we do a great disservice for generations to come.

Thank you very much.

[The prepared statement of Reverend Woodberry follows:]

THE WALL WE MUST BUILD

Justice First Campaign

Reverend Leo Woodberry

Kingdom Living Temple Church and New Alpha Community Development Corporation

Florence, SC

Honorable Members of the Congress US House of Representatives:

I would like to begin by thanking you for the opportunity to testify before the House Committee on Energy and its Commerce Subcommittee on Environment and Climate Change. This hearing entitled "Time for

Action: Addressing the Environmental and Economic Effects of Climate Change" is both urgent and timely.

I have been involved in environmental justice and climate change issues for over 25 years from a community, regional, national, and global levels. From April 2018 to August of 2018; I, along with more than 300 organizations and hundreds of individuals conducted a Justice First Tour through 12 southeastern states and 25 cities. During that tour, we were able to visit numerous communities that were impacted by environmental toxins and pollutions. Many of these communities were ravaged by devastating weather disasters, and intensified climate change. With the recent reports released by the Intergovernmental Panel on Climate Change (IPCC) and USA National Climate Assessment Report, there is irrefutable evidence that if we do not address the issues impacting our environment, climate, and economy, unprecedented devastation will occur. It is imperative that we construct a wall of mitigation, adaptation, and resilience to save the lives, property, communities, and the way of American life as we know it. We are told by the recent IPCC report (a conservative estimate) that we have less than 12 years to get on a serious path to reduce carbon emissions, or there will be little that we can do in the foreseeable future. As a resident of South Carolina, I have seen the increasing devastation being wrought throughout the south and throughout our country. In the lower ninth ward of New Orleans, Louisiana, communities are still struggling to recover from the impacts of Hurricane Katrina that occurred almost 14 years ago. During and after Hurricanes Florence and Michael in South Carolina, we saw widespread flooding and community evacuations (in some cases for the second time in two years). In Marion County, SC, just this past week, 248 residents were told that they had to elevate their homes or face the prospect of having no home owner's insurance when the next weather-related disaster strikes. Some of these residents are elderly, alone, and poor. Some of them are the people who harvested our crops and cleaned our homes, and whose employers never paid into Social Security for them. They, therefore, barely get by with the support of their friends and family with a SSI checks that typically is no more than $600 to $800 per month. Consider the prospect of the 90 year old woman whose home I visited in

Sellers, SC; she was told that she must elevate her home by 7 feet with no means to do so. I pray for her and others in anguish, who after the next hurricane comes, who if they have nowhere to turn, may transition in a nursing home, alone with a broken heart.

We must protect our American citizens and communities. We Must Build a Wall of mitigation, adaption and resilience that will create jobs in the areas of renewable energy, energy efficiency, and infrastructure. We must do this through lens that are both just and equitable: to those who will lose their jobs in the fossil fuel industries and those who have had their opportunities deferred and denied because polluting facilities were cited in their communities discouraging other businesses from locating there. We can only do these things that will strengthen us environmentally and economically by moving forward on a macro and micro level. By that I mean, we need community-based climate solutions, as well as community in place training and education, along with incentives for utilities, states and municipalities. Imagine the number of adaptation jobs that can be created as we elevate the homes of those residing in flood plains across our country. The people in urban areas desperately needing employment who can construct bios wells to mitigate flooding and clean up brownfields by using plant remediation. Federal support can also provide jobs to assist people in recovering from lost and damages while they clean out and restore drainage systems, particularly rural areas that for far too long have been left unattended.

We desperately need community based climate change solutions that can swiftly meet the environmental and climate change urgency of our nation. Standing shoulder to shoulder in a wall of solidarity, Americans can install demand side energy management devises to reduce our carbon emissions. Entrepreneur and employment opportunities can blossom when communities are allow to construct small scale on and off grid renewal energy devices and systems. Community in place training can provide the much needed education and training that will allow our community to become more resilient and recover from weather related disaster more quickly. Community in place training will provide the services that university and colleges will not be able to provide if they do not have tuition support.

The people of American need protection from many things. Climate change's weather related disasters do not differentiate between democrats or republicans, race or gender, geographic location, citizenship or immigrant status. Everyone within our boarders need to have a wall of protection as climate change become more intense and more frequent. I want to commend this august body for its proactive foresight, courage, love of this country and pray that your efforts will in result in definitive and fruitful results. Thank you for this wonderful opportunity to share the concerns of those who participated in the justice tour and everyone who continues to increase the effective capacity of addressing the Environmental and Economic Effects of Climate Change.

Mr. TONKO. Thank you, Reverend.

And now we will move to Mr. Barry K. Worthington. Mr. Worthington, you are recognized for 5 minutes.

STATEMENT OF BARRY WORTHINGTON

Mr. WORTHINGTON. Thank you, Chairman Tonko, Ranking Member Shimkus, and members of the Subcommittee on Environment and Climate Change. My name is Barry Worthington. I am the executive director of the United States Energy Association. I have been in this role for 30 years, and have another dozen years in the energy business.

The U.S. Energy Association has worked in transitional economies in developing countries for 25 years, over 25 years, with the U.S. Agency for International Development, and also with the Department of Energy, to expand the use of clean energy technology. Our members include energy production companies, energy efficiency companies, but also engineering, finance, legal, research, and consulting organizations. Our purpose is to convey information about the realities of global energy issues in the 21st Century.

We are not a lobbying organization. We are not an advocacy organization. We are an educational association both by function and IRS tax status. My intent today is to offer information and observations to you and to convey an offer that the U.S. Energy Association is available to be a resource for you and your staff as you begin to tackle the priorities of the 116th Congress.

The risks of climate change are real, and industrial activity around the globe is impacting the climate. Addressing climate change is a challenge for our country. It affects every world citizen. While the industry adjusts to climate change, it continues to ensure American citizens have access to increasingly safe, affordable, reliable, and clean energy, which we all do in this great country.

We are fortunate here. But we have between a billion and a billion-and-a-half global citizens with no access to commercial energy. Women in developing countries spend all day forging for sticks and animal dung to generate their cooking, lighting, and heating. This is dangerous. Burning firewood and animal dung indoors kills children. Indoor air pollution causes asthma and other health problems.

Access to energy, on the other hand, provides improved health, education, economic development, and allows mothers and fathers to spend more time with their family instead of scrounging around to find animal dung to burn in their—inside.

Central to energy access is lighting, for example. In developing countries, simple lighting reduces thefts, rapes, personal assaults, and other crimes. Access to energy paves the way for economic development in businesses such as simple cell phone charging enterprises, refrigeration for vaccines. Energy access improves people's lives.

And our members are volunteering their time to work with their counterparts in developing countries to share technology and management practices in the developing countries. And we are trying to do our part.

Our industry's challenge is to double the provision of energy services globally while reducing greenhouse gas emissions by 80 percent. Though there are 1 to 1.5 billion people with no access to energy, recognize there are also another 1.5 billion with inadequate access. And considering a global population growth of 2 billion leaves the energy industry to provide 5 billion more energy consumers access to energy services by mid-century.

Many of these consumers will utilize fossil fuels because they are domestic, abundant, and affordable. We should work harder towards helping them use high-efficiency/low-emissions technology. USEA has been doing this for 25 years.

And domestically we are expected to reduce greenhouse gas emissions by 80 percent. Our industry has undertaken a wide range of initiatives to reduce and avoid greenhouse gas emissions, and we are proud of our progress.

For example, electric power carbon dioxide emissions declined 28 percent from 2005 to 2017. Methane emissions declined 18.6 percent from 1990 to 2015, even though we increased domestic natural gas production by 50 percent. We think the solution to the dual challenges of climate change and global access to safe, reliable, and affordable and clean energy is technology. And an all-of-the-above approach is essential. This means all of the renewables as well as all of the traditional fuels, including nuclear and fossil fuels. We need to work harder towards assuring that fossil fuel utilization uses high-efficiency/low-emissions technology, including carbon capture and storage. Thank you, Mr. Chairman.

[The prepared statement of Mr. Worthington follows:]

TESTIMONY OF BARRY WORTHINGTON, EXECUTIVE DIRECTOR, UNITED STATES ENERGY ASSOCIATION, TO THE U.S. HOUSE OF REPRESENTATIVES,' COMMITTEE ON ENERGY AND COMMERCE, SUBCOMMITTEE ON ENVIRONMENT AND CLIMATE CHANGE

Chairman Tonko, Ranking Member Shimkus, and Members of the Subcommittee on Environment and Climate Change.

My name is Barry Worthington. I am the Executive Director of the United States Energy Association. I have been in this role for 30 years.

The U.S. Energy Association helps expand energy infrastructure in developing countries with the U.S. Agency for International Development (USAID) and drives policy and technical discussions with the U.S. Department of Energy to expand the use of clean energy technology around the world.

Through our membership, USEA also represents more than 100 companies and associations across the U.S. energy sector, from the largest Fortune 500 companies to small energy consulting firms. Our members include both energy production companies and energy efficiency companies, but also engineering, finance, legal, research and consulting organizations.

USEA's objective is to convey information about the realities of global energy issues in the 21st Century.

We do not lobby. We are not an advocacy group. We are an educational association both by function and IRS tax status.

Thank you for inviting me to appear before you today.

My intent is to offer you information and observations and to convey an offer to be a resource for you and your staff as you begin to tackle the priorities of the 116th Congress.

The risks of climate change are real, and industrial activity around the globe impacts climate. Addressing climate change is a challenge for our country. It affects every world citizen. It affects the energy industry trying to adapt and help energy end users adapt to changing weather patterns.

While the industry addresses the changing climate, it continues to do well to ensure American citizens access to increasingly safe, affordable, reliable, and clean energy.

We have more than 1 billion global citizens with no access to commercial energy. Women in developing countries spend all day foraging for sticks and animal dung to generate energy for cooking, lighting, and heating. This is dangerous. Burning firewood and dung indoors kills children, causes asthma, and other health problems.

Access to energy, on the other hand, provides improved health, education, economic development and it allows mothers and fathers to spend more time with their families.

Central to energy access is lighting. In developing countries, simple lighting reduces thefts, rapes, personal assaults, and other crimes. It paves the way for economic development. From businesses such as cell phone charging enterprises to refrigeration for vaccines, energy access improves human life.

Energy access and expansion in this country led to industrialization which has improved our health and welfare, built our economy and put the U.S. in the position to help the world achieve the same.

Our industry's challenge is to double the provision of energy services globally, while reducing greenhouse gas emissions by 80%.

Though there are 1.5 billion people with no access to energy, there are 1.5 billion other citizens with inadequate access. Considering a global population growth of two billion leaves the energy industry to provide 5 billion more energy consumers access by mid-century.

Many of these consumers will utilize fossil fuels because they are domestic, abundant, and affordable. We should work harder toward helping them use high efficiency/low emissions technologies. USEA has been doing this for over 25 years in over 50 countries.

And domestically we are expected to reduce greenhouse gas emissions by 80%. Our industry has undertaken a wide range of initiatives to reduce, avoid, or sequester greenhouse gas emissions. We are very proud of our progress.

For example, electric power carbon dioxide emissions declined 28% from 2005 to 2017. We expect this trend to continue.

Methane emissions from the natural gas industry declined by 18.6% declined from 1990 to 2015 even though U.S. natural gas production increased by more than 50%.

Since 2000, the energy industry has invested at least $120 billion in emissions-reducing technologies.

We think that the solution to the dual challenges of global climate change and global access to safe, reliable affordable and clean energy is through technology.

An "all of the above" approach is essential. This means all the renewables such as solar, wind, hydro and geothermal, as well as traditional fuels and technology such as nuclear and all the fossil fuels. We need to work towards assuring that fossil fuel utilization uses high efficiency/low emissions technology including carbon capture and storage.

Americans lead the world in innovation and we can complete the energy revolution that began in earnest a decade ago. Increased domestic energy production has resulted in lower emissions of carbon dioxide and pollutants while lowering costs to consumers.

Mr. Chairman, I have two more points.

The first is I met recently with the state and national leaders involved with the Low-Income Home Energy Assistance Program known as LIHEAP. They have convinced me that we do not totally appreciate the impact of energy costs on low-income consumers.

I know our industry has done a great job of lowering energy price. This is indisputable. But having low energy prices does not mean that a large number of Americans will not have trouble paying their energy bills.

While our economy is strong, employment numbers are up, and wages are increasing, we still have a sizeable percentage of our population that cannot afford to have their energy bills increase.

I hope you will consider this in your deliberations on how we respond to climate change. Thank you for your kind attention.

Mr. TONKO. Thank you very much, Mr. Worthington.

And finally, from the BlueGreen Alliance, Mr. Michael Williams.

You are recognized for 5 minutes.

STATEMENT OF MICHAEL WILLIAMS

Mr. WILLIAMS. Thank you, Chairman Tonko, Republican leader Shimkus, distinguished members of the committee. I am honored to be here alongside my fellow panelists and with you all as we strive to find common comprehensive solutions.

As the chairman noted, my name is Mike Williams. I am the deputy director of the BlueGreen Alliance, a national partnership of labor unions and environmental organizations. BlueGreen Alliance unites America's largest labor unions and its most influential environmental organizations to solve today's environmental challenges in ways that create and maintain quality jobs and build a stronger, fairer economy.

We believe that Americans don't have to choose between a good job and a clean environment or a safe climate. We can and we must have both.

The world's leading scientific organizations have been unambiguous that climate change is a dire and urgent threat. And we need comprehensive action and solutions to rapidly drive emissions down now. I am heartened by the common commitment to action I am hearing today.

Our communities bear the burden of climate change in wildfires, hurricanes, heat waves, droughts, and sea level rise it spawns. At the same time, our Nation is struggling with deep and crippling economic inequality. The majority of American families are less able to deal with these problems as their wages have fallen and their economic mobility and power in the workplace has declined.

For too long the debate on the economic impact of climate action has been framed as either disaster or miracle, yet neither aligns with the complicated realities in which American workers live. This flawed debate has prevented us from addressing climate change at a level commensurate with the size of the challenge. The driving forces behind the challenges of climate change and inequality are intertwined, and we must tackle them together as equal priorities and place good jobs and working families at the center of a massive economic transformation.

Thankfully, we are starting to see examples across the country of the kinds of solutions needed to achieve this outcome and justice for all Americans. Take Buy Clean California, a landmark law that requires State agencies to consider the embedded carbon emissions of industrial products. This law will reduce emissions globally, while also leveling the playing field for domestic manufacturers who are investing in clean, efficient manufacturing technologies and processes.

Or in the State of Illinois, where the Future Energy Jobs Act provides sweeping changes to boost renewable energy and energy efficiency while protecting the jobs of workers at current energy generation facilities in the State, including existing nuclear power plants, and establishing standards for the solar industry to use a skilled and qualified workforce.

Finally, critical Federal efforts, like America's landmark fuel economy and greenhouse gas standards for cars and trucks, drive investment, innovation, and job growth. Our research finds more than 1,200 U.S. factories and engineering facilities in 48 States, and 288,000 American workers, building technologies that reduce pollution and improve fuel economy for today's innovative vehicles.

As significant transformation is needed to truly address climate change and inequality at the speed and scale demanded by the scientific reality and the urgent needs of our communities, it will require bold ideas and a guarantee that no worker or community is left behind. And instead of leaking jobs and pollutions overseas, we invest in our industries and our people here.

This is a big task. But I cannot stress firmly enough that no solution to climate change or inequality will be complete if Congress does not move forward with an ambitious plan to rebuild and transform America's infrastructure so that it is ready for the significant transformation we need to tackle climate change. This plan must address all aspects of our infrastructure needs, from strengthening the electric grid and modernizing our water systems to reducing methane leaks in the natural gas distribution sector, improving surface transportation, investing in natural infrastructure, and making our schools, hospitals, and other buildings safer, healthier, and more energy efficient.

These investments can reduce air and water pollution and make our communities more resilient to the impacts of climate change. They will also create millions of good jobs. But we have to make sure we tackle this challenge the right way.

This means ensuring all products are subject to Buy America and Davis-Bacon; using project labor agreements and community benefit agreements, and local hire provisions; prioritizing the use of the most efficient, resilient, and cleanest materials and products; enhancing workforce training and development programs; increasing pathways to economic opportunities for communities and local workers, especially people of color and low-income communities; and prioritizing public funding and financing.

Repairing America's infrastructure systems should be a bipartisan legislative priority for the 116th Congress.

In closing, I want to reiterate that tackling the crisis of climate change, if done right, is a significant opportunity to ensure a more equitable society, increase U.S. global competitiveness, and create quality, family-sustaining jobs across the country.

We look forward to working with this committee as you move forward with your agenda for the 116th Congress. Thank you again for the opportunity to testify.

[The prepared statement of Mr. Williams follows:]

WRITTEN TESTIMONY, MICHAEL WILLIAMS, DEPUTY DIRECTOR, BLUEGREEN ALLIANCE, BEFORE THE 116TH UNITED STATES CONGRESS, HOUSE COMMITTEE ON ENERGY AND COMMERCE, SUBCOMMITTEE ON ENVIRONMENT AND CLIMATE CHANGE

Rayburn House Office Building, Room 2123
Wednesday, February 6, 2019

Thank you Chairman Tonko, Ranking Member Shimkus, and distinguished members of the committee. My name is Mike Williams, and I am the Deputy Director of the BlueGreen Alliance, a national partnership of labor unions and environmental organizations. On behalf of my organization, our partners, and the millions of members and supporters they represent, I want to thank you for convening this hearing today.

The BlueGreen Alliance unites America's largest labor unions and its most influential environmental organizations to solve today's environmental challenges in ways that create and maintain quality jobs and build a stronger, fairer economy. Our partnership is firm in its belief that Americans don't have to choose between a good job and a clean environment—or a safe climate—we can and must have both.

The world's leading scientific organizations have been unambiguous that climate change is a dire and urgent threat and that the longer we delay, the stronger the action required.

Over the last decade, we have witnessed the worsening impacts climate change is having on our communities. America's workers are often on the frontlines of these impacts—not only feeling the effects of drought or extreme weather in their own communities, but responding to these crises.

The 2017 historic hurricane season laid waste to Puerto Rico, the U.S. Virgin Islands, Texas, and Florida, in some areas destroying critical infrastructure systems, plunging millions of Americans into darkness, and further aggravating an already desperate need for safe water. Immediately, brothers and sisters of some of the nation's largest labor unions worked to provide much-needed relief to communities in distress.

The American Federation of Teachers (AFT) launched Operation Agua to provide safe drinking water to families across Puerto Rico.[68] When flooding caused by Hurricane Harvey forced hundreds of thousands of people from their homes in Texas, members of United Steelworkers (USW) were among those to brave the rising waters to rescue individuals and families stuck in dangerous situations.[69]

Members of the Utility Workers Union of America (UWUA) were among the first to spring into action after Hurricanes Harvey and Irma left thousands of residents in Texas and Florida without electricity. Members

---

[68] American Federation of Teachers, "Operation Agua." Available online: https://www.aft.org/operation-agua.

[69] USW, "Not even a hurricane can stop USW." September 12, 2017. Available online: https://www.youtube.com/watch?v=-24WHNN5fS8.

from across the country loaded up their trucks to help get the lights back on in impacted communities.[70]

The United Association of Plumbers and Pipefitters (UA), International Association of Sheet Metal, Air, Rail, and Transportation Workers (SMART), and the Service Employees International Union (SEIU) launched disaster relief funds and collected and distributed supplies throughout the nation to those in need. Volunteers across the U.S. and Canada collected and delivered boxes, fans, dehumidifiers, bleach, food, diapers, and other items to members in affected communities.[71,72]

America's workers were on the ground, helping to provide critical services to those in need, volunteering their time and money to those who had lost everything.

These impacts are only going to continue and worsen if we fail to adequately address the crisis of climate change. Our communities bear the burden of climate change and the wildfires, hurricanes, heat waves, droughts and sea level rise it spawns. The majority of American families are less able to deal with these problems as their wages have fallen and their economic mobility and power in the workplace has declined. As Chris Shelton, President of the Communications Workers of America (CWA), and Mary Kay Henry, President of SEIU, stated, "These forces wreak havoc on our lives and ability to provide for our families due to increased health-related calamities, financial stress and property damage. This is on top of the realities that working families already face in an economy that isn't working for us."[73]

At the same time, our nation is struggling with deep and crippling economic inequality. According to the Economic Policy Institute, "the bottom 90 percent of the American workforce has seen their pay shrink

[70] UWUA, "UWUA Locals Spring into Action in the Wake of Hurricanes Irma and Harvey." September 26, 2017. Available online: https://uwua.net/news/uwua-locals-spring-action-wake-hurricanes-irma-harvey/.

[71] UA, "UA reaches out to local community to aid those effected by Hurricane Harvey." April 16, 2018. Available online: https://www.youtube.com/watch?v=Yq-zNKfxmxM.

[72] SMART, "Mobilizing for Hurricane Relief." September 29, 2017. Available online: https://www.youtube.com/watch?v=Yq-zNKfxmxM.

[73] SEIU, CWA, "America's working families demand solutions, not cop-outs on climate." April 27, 2017. Available online: https://medium.com/@ChrisSheltonCWA/americas-working-families-demand-solutions-not-cop-outs-on-climate-db2956026468.

radically as a share of total income," from 58 percent in 1979 to 47 percent in 2015. That is almost $11,000 per household, or $1.35 trillion in additional labor income. There is direct correlation with the decrease of worker power over this time, as the share of workers in a union fell from 24 percent in 1979 to under 11 percent now.[74]

For too long, the debate on the economic impact of climate action has been framed as either disaster or miracle, yet neither aligns with the complicated realities in which American workers live. This flawed debate has prevented us from addressing climate change at a level commensurate with the size of the challenge. The driving forces behind the challenges of climate change and inequality are intertwined and we must tackle them together as equal priorities. We will only overcome these twin challenges if good jobs and working families are at the center of a massive economic transformation.

Limiting global warming to the extent required by science will, according to the Intergovernmental Panel on Climate Change (IPCC), "require rapid, far-reaching and unprecedented changes in all aspects of society," and "could go hand in hand with ensuring a more sustainable and equitable society."[75] This transformation must happen at the speed and scale demanded by scientific reality and the urgent needs of our communities. If we do it right, we can not only avoid the worst impacts of climate change, but create quality, family-sustaining jobs and ensure a more equitable society.

Thankfully, we are starting to see examples across the country of the kinds of solutions needed to achieve this outcome.

---

[74] Economic Policy Institute, "What labor market changes have generated inequality and wage suppression?" December 12, 2018. Available online: https://www.epi.org/publication/what-labor-market-changes-have-generated-inequality-and-wage-suppression-employer-power-is-significant-but-largely-constant-whereas-workers-power-has-been-eroded-by-policy-actions/.

[75] IPCC, Summary for Policymakers of IPCC Special Report on Global Warming of 1.5°C approved by governments.October 8, 2018. Available online: https://www.ipcc.ch/2018/10/08/summary-for-policymakers-of-ipcc-special-report-on-global-warming-of-1-5c-approved-by-governments/.

First, in California, a landmark law signed by Governor Brown in 2017 known as "Buy Clean California" is beginning to be implemented.[76] This bill, which passed the California state legislature with bipartisan support, is the world's first legislative effort to address imported carbon emissions and will require state agencies to consider the embedded carbon emissions of industrial products—like steel and glass—when contracting for state-funded infrastructure projects. The law will reduce emissions associated with significant infrastructure projects while leveling the playing field for manufacturers who are investing in clean, efficient manufacturing technologies and processes and protecting good American jobs at the same time.

Second, off the coast of Rhode Island, America's first offshore wind farm, the Block Island Wind Farm, has been fully operational since the end of 2016. This project not only reduces greenhouse gas emissions by generating carbon-free electricity, but this comparatively small demonstration project created over 300 jobs in Rhode Island alone for local unionized craftsmen in ten different Building Trades locals working for 30 unionized contractors and subcontractors. This was thanks, in large part, to the Project Labor Agreement (PLA) in place.[77]

This innovative new project, and even bigger projects in development in states like Connecticut, Maryland, Massachusetts, New Jersey, New York, and Virginia, has the potential to dramatically expand quality job creation in a relatively untapped sector. Estimates put job creation potential off the Atlantic coast alone at between 133,000 and 212,000 per year in the United States.[78] The National Renewable Energy Laboratory (NREL) cites that the Atlantic states could create $200 billion in new economic activity,

---

[76] BlueGreen Alliance, "Buy Clean California Act Claims Down on Imported Carbon Emissions." October 16, 2017. Available online: https://www.bluegreenalliance.org/the-latest/buy-clean-california-act-clamps-down-on-imported-carbon-emissions/.

[77] National Wildlife Federation, "The Unifying Appeal of Offfshore Wind Power." June 16, 2017. Available online: https://blog.nwf.org/2017/06/the-unifying-appeal-of-offshore-wind-power/.

[78] Oceana, "Offshore Wind Power Report Key Findings." September 28, 2010. Available online: https://usa.oceana.org/offshore-wind-report-key-findings.

as well as over 43,000 high-paying, permanent jobs, just by developing 54 gigawatts of their 1,283 gigawatt offshore wind energy potential.[79]

In order to truly capture the full benefits and potential of these projects, it is critical that they are built and operated by skilled and properly trained workers that are paid family-sustaining wages, with Project Labor Agreements in place, and with products and components throughout the supply chain manufactured here at home.

A third example comes from the state of Illinois, where the Future Energy Jobs Act was signed into law in December 2016. The bill provides sweeping changes to the state's Renewable Portfolio Standard (RPS) and increases state efforts to boost energy efficiency, while protecting the jobs of workers at current energy generation facilities in the state, including existing nuclear power plants, and establishing standards for the solar industry to use a skilled and qualified workforce. The labor and the environmental movements worked hand-in-hand to push this bill over the finish line.[80]

We are also seeing great examples of how innovating, manufacturing and installing the clean economy—particularly through clean vehicles and energy efficiency—is protecting the environment while creating quality jobs. Across the country, millions of jobs in manufacturing depend on continued American leadership in clean vehicle technology, including building innovative, far more efficient cars, SUVs, and trucks. Our research finds more than 1,200 U.S. factories and engineering facilities in 48 states—and 288,000 American workers—building technologies that reduce pollution and improve fuel economy for today's innovative vehicles. Critical federal efforts—like America's landmark fuel economy and greenhouse gas standards for cars and trucks—drive investment,

---

[79] NREL, Large-Scale Offshore Wind Power in the United States. September 2010. Available online: https://www.nrel.gov/docs/fy10osti/49229.pdf.

[80] Union of Concerned Scientists, "A Huge Success in Illinois: Future Energy Jobs Bill Signed Into Law." December 8, 2019. Available online: https://blog.ucsusa.org/jessica-collingsworth/big-win-illinois-energy.

innovation, job growth, and help position the domestic industry as a global leader.[81]

Energy efficiency improvements for public buildings and manufacturers support existing jobs, increase the quality of jobs created, and reduce carbon pollution. Local communities can also capture the benefits of energy efficiency retrofits—including lower utility bills, improved tenant health, and increased economic development. In California, new rules adopted by the California Public Utilities Commission (CPUC) will ensure that when utility incentives are used for certain efficiency upgrades, the workers installing the equipment are properly trained, and that efficiency programs create pathways to employment for disadvantaged workers.[82] We view this as a significant development.

Furthermore, identifying the supply chain and moving policies to increase domestic production of energy efficiency retrofit products can also help to create quality manufacturing jobs. The BlueGreen Alliance Foundation's Building Clean program works to identify these products and advance energy efficiency retrofits, particularly for multi-family housing.[83]

These are the kind of worker- and community-centric solutions that will equitably address the climate crises and create quality jobs. These are models for the federal level because we have yet to adequately take action commensurate with the challenges of climate change and inequality.

A significant transformation in public policy is needed to truly address climate change and inequality at the speed and scale demanded by scientific reality and the urgent needs of our communities. It will require

---

[81] BlueGreen Alliance and the Natural Resources Defense Council, Supplying Ingenuity II: U.S. Suppliers of Key Clean, Fuel-Efficient Vehicle Technologies. May 23, 2017. Available online: https://www.bluegreenalliance.org/resources/supplying-ingenuity-ii-u-s-suppliers-of-key-clean-fuel-efficient-vehicle-technologies/.

[82] International Brotherhood of Electrical Workers Ninth District, Western States Council of the International Association of Sheet Metal, Air, Rail and Transportation, and CA State Pipe Trades Council, "California Public Utility Commission Agrees: A Skilled, Trained, and Diverse Workforce is Key to Achieving Efficiency Goals." January 11, 2019. Available online: https://www.bluegreenalliance.org/resources/california-public-utility-commission-agrees-a-skilled-trained-and-diverse-workforce-is-the-key-to-achieving-efficiency-goals/.

[83] BlueGreen Alliance Foundation, "Building Clean." Available online: http://www.building clean.org.

bold ideas to create and retain millions of high-quality jobs, including a significant commitment to increasing union density across the country, the right to organize in clean technology sectors and throughout the economy, and mandatory labor standards.

It will require a revitalization of U.S. manufacturing with a national commitment to leading the world in building the clean technologies and materials of the future in the U.S. and doing so with clean, efficient industrial processes and products. A new generation of advanced manufacturing growth can help rebuild the middle class and at the same time decarbonize our industrial sector, which has received relatively less attention from policymakers while becoming a large source of recent U.S. emission increases.[84] It can also help U.S. manufacturers capture market share in in a changing global economy where clean products and processes will represent an essential competitive advantage.

It will require that we dramatically increase the capacity of our public sector, our health care system, and the community based non-profit sector to prepare for and respond to the demands of a changing climate. And it will go hand in hand with ensuring that no worker or community is left behind and that all workers and communities have access to safe, high-quality, union jobs—as well as clean air and water.

To truly be transformative, this economic mobilization must tackle the economic and racial inequity that exposes low-income workers, communities of color, and others to low wages, toxic pollution, and climate threats. We must inject justice into our nation's economy by ensuring that the economic and environmental benefits of this transformation support first and foremost those workers and communities that have been hardest hit by the unjust status quo.

This is a big task. But there are concrete steps this Congress and this Committee can take to put us on this path, some of which have been have described above. Bold ideas have been put forth—and more will be coming—but I cannot stress firmly enough that no solution to climate

---

[84] New York Times,"U.S. Carbon Emissions Surged in 2018 Even as Coal Plants Closed." January 8, 2019. Available online: https://www.nytimes.com/2019/01/08/climate/greenhouse-gas-emissions-increase.html.

change or inequality will be complete if Congress does not move forward with an ambitious plan to rebuild and transform America's infrastructure so that it is ready for the significant transformation we need.

American infrastructure systems are in dire need of repair and modernization. The American Society of Civil Engineers (ASCE)'s latest 2017 Report Card for America's Infrastructure gave the nation's infrastructure a grade of "D+," and estimated that getting to a grade of "B" would require significant federal investment over the next 10 years.[85]

Congress must move forward a plan to meet this need, including boosting revenue to support the public sectors' ability to meet the challenge. A robust and ambitious infrastructure plan should address all aspects of our infrastructure needs, from strengthening the electric grid and modernizing our water systems to reducing methane leaks in the natural gas distribution sector, improving surface transportation, and making our schools, hospitals, and other buildings safer, healthier, and more energy efficient.

Just rebuilding our schools can have an impact that reverberates across the economy. As James Boland, President of the International Union of Bricklayers and Allied Craftworkers (BAC), noted, "It is critically important that we rebuild our crumbling school facilities now so that our students are able to attend schools that are conducive to learning and that our teachers can do their important work in safe and healthy environment."[86]

All of these investments can reduce air and water pollution—including the emissions driving climate change—and make our communities more resilient to the impacts of climate change. Natural infrastructure can also improve climate resilience through natural defenses that act as carbon sinks like wetlands, forests, dunes, and grasslands.

---

[85] ASCE, 2017 Infrastructure Report Card. Available online: https://www.infrastructure reportcard.org.

[86] Press release from Congressman Bobby Scott, "House and Senate Democrats Unveil Proposal to Invest More than $100 Billion in America's Public Schools." January 30, 2019. Available online: https://bobbyscott.house.gov/media-center/press-releases/house-and-senate-demo-crats-unveil-proposal-to-invest-more-than-100.

Making these smart investments will also pay dividends for workers and our economy. Our research has found that investing an estimated $2.2 trillion in these sectors to improve them from a "D+" grade overall to a "B" grade has the potential to support or create an additional 14.5 million job-years across the U.S. economy and add a cumulative $1.66 trillion to Gross Domestic Product (GDP) over 10 years, versus a business as-usual approach.[87]

We will accrue these benefits only if we tackle this challenge the right way. To ensure we maximize the benefits of our infrastructure investments for communities, the environment, and workers, any infrastructure package must:

- Ensure all projects built with public resources are subject to Buy America standards that maximize the return to taxpayers, workers and the American economy;
- Enforce Davis-Bacon provisions that ensure workers are paid prevailing wages;
- Utilize project labor agreements (PLAs), community benefit agreements, local hire, and other provisions and practices that prioritize improving training, working conditions, and project benefits;
- Drive forward-looking planning and investments that meet environmental standards and build resilient infrastructure systems and communities;
- Prioritize use of the most efficient, resilient, and cleanest materials and products with the lowest carbon and toxicity footprints;
- Enhance workforce training and development programs to expand the number of skilled workers in new and existing industries;
- Increase pathways to economic opportunities for communities and local workers, especially for people of color and low-income communities;

[87] BlueGreen Alliance, *Making the Grade 2.0: Investing in America's INfrastructure to Create Quality Jobs and Protect the Environment*. September 7 2017. Available online: https://www.bluegreenalliance.org/makingthegrade2.

- Prioritize public funding and financing for infrastructure investment. All financing methods should be held to strong public interest standards.

Repairing America's infrastructure systems is both urgently needed and an enormous opportunity; it should be a bipartisan legislative priority in the 116th Congress. We urge you to develop and put forward an infrastructure package that meets the above criteria.

In closing, I want to reiterate that tackling the crisis of climate change—if done right—is a significant opportunity to ensure a more equitable society, increase U.S. global competitiveness, and create quality, family-sustaining jobs across the country. Given the scale of the problem, numerous solutions will be needed. We encourage the committee to use every tool at its disposal to make progress now.

We look forward to working with this Committee as you move forward your agenda for the 116th Congress. Thank you again for the opportunity to testify today.

Mr. TONKO. I thank you, Mr. Williams, and your fellow panelists, who have provided great information.

So that concludes our opening statements. We will now move to Member questions. Each Member will have 5 minutes to ask questions of our witnesses. I will start by recognizing myself for 5 minutes.

The United States emits around 6.5 billion metric tons of greenhouse gas each and every year. That pollution will outlast us by decades, and even centuries. As is clear from testimony, Americans are already feeling the effects of climate change, but most of the people in this room will be long gone when the worst consequences hit. The decisions we make today will determine the conditions for generations not yet born.

Dr. Ekwurzel, I would like you to expand upon why it is so important that we start drastically reducing emissions now.

Dr. EKWURZEL. Thank you, Chairman Tonko.

Essentially what you said is correct, that for 20 percent of the carbon dioxide emissions it could be trapping heat day-in, day-out for centuries. And also methane, nitrous oxide, these are the very important pollutants to get out of the atmosphere. In part, because you may have noticed that coastal properties is one of the big sectors for damage. And if you reduce emissions you can take over a 20 percent bite out of that. And it is because the legacy of sea-level rise has already been baked in with the historical emissions of heat—trapping gases into our atmosphere.

So think about what else we have baked in. It is very important to reduce emissions now so we have a chance at taking a 60 percent bite out of damages and extreme heat mortality in the labor sector, 50 to 60 percent. It is critical for saving lives to reduce emissions as soon as possible. Delay is super costly.

Mr. TONKO. And the difference between a high-emissions or business-as-usual scenario compared to a low-emissions one, what basically is that difference?

Dr. EKWURZEL. So, for example, in damage to the U.S. economy, the loss of labor cost, the range could be $20 to $200 billion per year by the year 2090.

If we went on the low-emissions pathway, we could take nearly a 60 percent bite out of that, or 50 to 60 percent. And that doesn't include adaptation. If we add adaptation in the mix, we can lower the costs immensely.

What we see is, in general, a very tight relationship with each global average surface temperature increase, a bigger bite out of the U.S. percentage GDP.

Now, Ranking Member Walden mentioned some of the costs to transition to a clean energy economy. You compare that against some of these annual costs, you start realizing that an investment in reducing emissions is a very good investment.

Mr. TONKO. Thank you.

And, Mr. Duke, you have done a lot of work on decarbonization strategies. I, for one, believe we cannot take solutions off the table at this point. I hope today we can hear about the merits of many different options.

Given all the potential pathways to decarbonize our economy, at this stage in the process how would you recommend Congress approach this challenge?

Mr. DUKE. Thank you, Chairman. I would start on two tracks to address this challenge, starting with the easiest part first. And that would include at least doubling clean energy and clean solution research and development investment. And I appreciate the bipartisan move in that direction over the last year or two.

And at the same time, in the near term it is possible to do quite a bit of harvesting of low-hanging fruit. That includes things like measures to cut energy waste, to scale renewables even faster because they do need to go even faster than today's pace. We need to modernize the electricity grid, as has been noted. And do things that save consumers money, and cut energy waste, and build on what the States are already doing.

At the same time, we need to go the next step. And the next step on a second track would be putting in place comprehensive policies that start with a price on carbon sufficient to put us on that path to net-zero greenhouse gas emissions by mid-century. And we need to do this in a way that ensures that all communities benefit equitably and that we're investing the resulting revenue in a smart way. This will create broad-based economic incentives that help our entrepreneurs and innovators scale up and bring down costs yet further and create that global momentum that we need.

Mr. TONKO. Thank you very much.

I share the sentiment that we need to make progress now while we can, while developing our comprehensive economywide solution.

I mentioned before that it has been a decade since the House last seriously attempted to address climate change. What has changed over the past 10 years that indicates that this time it can be different, Mr. Duke?

Mr. DUKE. Thank you for the question. There is quite a bit on the technology front that is worth just briefly summarizing.

We have got all kinds of cost-effective solutions today, from wind and solar to energy efficiency. And electric vehicles are even cost effective for some drivers in high-mileage applications, like taxi drivers. You see them even here in DC.

And you have got demand flexibility solution as well that are helping with the intermittency of some renewables.

Down the line we see all kinds of things coming soon, like emerging technologies that electrify heating buildings through heat pumps, and electric vehicles that are cheap enough to compete on first cost with internal combustion engines, and dominate in terms of life cycle costs, will be available by many estimates within 5 years.

And so this kind of technology solution set is a game changer and making it easier to act to cut pollution today.

On the policy side, we have also learned a lot. And I think it is worth noting that pricing pollution clearly works. And what we have seen, in fact, is that countries that have done this, for example the European Union or our own States in the Northeast or California, have routinely seen that innovation means that the cost of the tradable permits under a cap-and-trade system is much lower than they initially anticipated.

And so we should think about that as a lesson to create investor certainty when we have these kinds of programs. We might want to add a price floor on those kinds of mechanisms. And in general we need to ratchet up standards regularly for things like efficiency so we don't lose momentum on fuel economy or appliance efficiency. And we need to stretch incentives further with competitive mechanisms like clean electricity standards.

Mr. TONKO. Thank you. Thank you, Mr. Duke.

And I now recognize Representative Shimkus as the Republican leader of this subcommittee for 5 minutes to ask questions.

Mr. SHIMKUS. Thank you, Mr. Chairman. I'm going to turn my questions to Mr. Worthington.

You state that the challenge for the energy industry is to double the provision of energy services globally while reducing greenhouse gas emissions by 80 percent. Can you break this down for me? What is driving the increase in global energy demand? And why are fossil fuels projected to remain the dominant source for energy globally?

Mr. WORTHINGTON. Thank you, sir, for that question.

Driving demand is multifold. It is a 2 billion population increase by the middle part of the century. It is providing access to energy for a billion to 1.5 billion people who don't have it now. This is captured in the United

Nations Sustainability Goal Number 7. And it is increasing the availability of energy to those citizens today who don't have reliable, affordable access to energy.

There are countries in, for example, in Africa and Asia where electricity might be available 3 to 4 hours a day. And that just renders an economy helpless. You can't operate industrial facilities with electricity only being available 3 or 4 hours a day.

So those are the drivers of demand.

On the production side, you know, we work in dozens and dozens of countries. We are in touch daily with the people who operate energy systems in other countries. And in China, India, Indonesia, Vietnam, South Africa, Colombia, so on and so forth, they all tell us they have every intention of continuing to use their domestic fossil energy resources because they are domestic, they don't have to be imported, they are abundant, and they are affordable.

And I have had business people tell me, "Don't pay attention to what our government leaders say about us, we are going to use fossil"——

Mr. SHIMKUS. OK, wind this up because I have got a couple more questions for you, so.

Mr. WORTHINGTON. OK. "We are going to continue to use fossil energy."

Mr. SHIMKUS. Thank you. What is the scale of transition that would have to take place to reduce energy system emissions by 80 percent?

Mr. WORTHINGTON. Well, we would have to deploy every type of low-carbon/no-carbon technology that is possible. This truly becomes an all-of-the-above, and recognizing that countries are going to continue using fossil fuels.

Mr. SHIMKUS. Well, let me ask this: Can the world do that with existing technology? Can they do it now?

Mr. WORTHINGTON. We can't do it today, no. We need technology advancement all across the board, advanced nuclear systems, better energy storage, better renewables, and carbon capturing and the like.

Mr. SHIMKUS. Which I think it speaks to the research and development equation that a lot of you have supported. Because we can't do it now, but with R&D and continued dollars we may be able to get there eventually. Correct?

Mr. WORTHINGTON. If we can put a man on the moon, we can solve the climate problem.

Mr. SHIMKUS. My friend McNerney would say it is an engineering problem, right? He is right there. He is a Californian, so.

That is right. You are going to be a long time before you get to ask questions.

Some climate change proponents want to move fully away from fossil energy. Is your experience in this reasonable?

Mr. WORTHINGTON. Impossible.

Mr. SHIMKUS. Is there another way at the problem where the benefits of affordable energy help us actually address climate risk?

Mr. WORTHINGTON. Yes. By deploying technologies that reduce the CO2 output from fossil energy: high-efficiency/low-emissions technologies.

Mr. SHIMKUS. Yes, I think you weaved a great story in your opening statement. I think we all know people who are in different aspects, maybe in the mission field in underdeveloped countries. And I think understanding—and the Reverend is here—and we are concerned about our brother, and we are supposed to be our brother's keeper, bringing electricity to underdeveloped countries helps their livelihood, helps them develop, helps them or their State.

So that is part of the whole discussion as we deal with this, not just as a United States solution but as a solution that will affect the entire world.

You are the current chairman of the Committee on Cleaner Electricity Production for Fossil Fuels for the United Nations Economic Commission for Europe and a member of the Sustainable Energy Committee for the U.N. Commission. How would you describe the role of fossil fuels in meeting U.N. sustainability goals?

Mr. WORTHINGTON. The U.N. Sustainability Goal Number 7 is energy access. And the use of traditional fuels all around the world are critical to achieving that goal.

Mr. SHIMKUS. Thank you.

Mr. Chairman, I will give you the 2 seconds left.

Mr. TONKO. Thank you. Thank you. The gentleman yields back.

Now the Chair recognizes Representative Pallone, full committee chairman, for 5 minutes to ask questions.

Mr. PALLONE. Thank you. I just wanted to emphasize, Mr. Chairman, the priority for our committee in addressing climate change. And to that end, I do believe we can work together, and it will strengthen the economy and create more good-paying jobs in addition to protecting the environment through investments in clean energy and resilient infrastructure.

So I want to start with Dr. Ekwurzel. What does the Fourth National Climate Assessment say about the anticipated effects of climate change on our Nation's infrastructure?

Dr. EKWURZEL. It is we do need to build a more resilient infrastructure in the United States to deal with the earlier snow melt in the western mountains, and providing water that is escaping out of water sheds that we could instead harness for water resources, fighting wildfires, and other aspects. We need to upgrade our 20th century infrastructure to deal with the 21st century climate impacts. And that is a wise investment.

Mr. PALLONE. Well, I believe very strongly that if we are going to do something on a bipartisan basis to address climate change that a major infrastructure bill and putting provisions in that bill will probably be the thing that we can most easily do on a—maybe "easy" is not the word, but that we can most likely do on a bipartisan basis and get President Trump to sign.

But do we have the tools to address this? In other words, how do we make—how can we build and repair infrastructure in ways that reduce pollution? Give us some ideas and how feasible that is.

Dr. EKWURZEL. Sure. When you take climate change risks into account, you end up having solutions, such as on the coastal areas, of nature-based solutions that are more resilient to the different types of hazards that climate-induced extreme events throw your way, and they suck up more carbon. So that is important and helps reduce emissions.

However, if we do other types of infrastructure decisions that do not take into account the risks or the increased emissions that may result, we could make it, you know, have maladaptive options. We have to learn as we go and start as soon as possible.

Mr. PALLONE. You are saying that we have to be careful if we do a major infrastructure bill that we actually, you know, build in these provisions that will help address climate change, otherwise it might make it worse?

Dr. EKWURZEL. Yes. And we have a lot of folks that are stepping up with lots of interesting designs once these incentives are unrolled.

Mr. PALLONE. All right, let me ask Mr. Williams about job opportunities associated with expanding clean and renewable energy. How do we ensure that, you know, that what we do with clean and renewable actually creates jobs and supports and strengthens the middle class?

Mr. WILLIAMS. Sure. I appreciate the question, Mr. Chairman.

Mr. PALLONE. And, again, by reference to infrastructure, if you could.

Mr. WILLIAMS. Yes, absolutely. Infrastructure is a phenomenal way to do that. So direct investment in infrastructure across systems, especially in the electricity, in the energy grid, so both the deployment of energy for heating and transportation, as well as electricity. So directly investing in that area of infrastructure is incredibly important. But doing so in a way that advances strong labor standards or incorporates strong labor standards.

So what we think of as basic items like prevailing wage standards, buy American, standards that make sure that when direct Federal investment goes into these projects that we are ensuring that high quality——

Mr. PALLONE. Give me some examples. You mentioned the electricity grid. What else? What about pipelines? What about, you know, electric vehicles?

Mr. WILLIAMS. Absolutely. So, for us to deploy electric vehicles across the country, we will need a massive upgrade in electric vehicle infrastructure, charging stations, so on and so forth, across the country. That is an incredibly important one.

You mentioned pipelines. Water infrastructure is absolutely critical. We often don't realize the amount of energy we use pumping water through our system. And when you are leaking water out of leaky old systems, you are losing energy and increasing pollution. So, simply by upgrading water infrastructure systems, we actually would save energy and reduce pollution. And all of that could and should be high-quality job creation.

Mr. PALLONE. And I, you know, I hear in New Jersey there are all kinds of pipelines being built. And, you know, different people are for it or against it. But I keep reminding them that, rather than focusing on new pipelines, why not focus on repairing existing, even for the natural gas? I mean, you can do a lot with maintenance and repair there that makes a difference in terms of climate change too, right? It is not just water, it is also natural gas and——

Mr. WILLIAMS. Yes. So we have long had a campaign for a number of years on repairing and replacing natural gas distribution systems, the distribution systems under the city that deliver natural gas to homes and businesses so that they can heat properly. And those systems are old and they are leaky and they can be dangerous, so repairing them should be an absolute priority, not only because of the pollution that would save but the high-quality job creation, as well as the safety concerns.

Mr. PALLONE. Thank you. Thank you, Mr. Chairman.

Mr. TONKO. The gentleman yields back.

The Chair now recognizes Representative Walden, full committee Republican leader, for 5 minutes.

Mr. WALDEN. Thank you, Mr. Chairman. And thanks——

Mr. TONKO. Five minutes to ask questions.

Mr. WALDEN. Thank you. And thanks again for having this hearing. I want to thank our panelists. Your testimony really will inform our work, and we appreciate it.

And, Mr. Williams, I appreciate your comments about, I believe you talked about the grid and improving drinking water supplies and things of that nature. I think we did 12 hearings in the last 2 years on grid adequacy, security. As we look to integrate new resources onto the grid, we have got to make sure it will handle the new renewables and the spikes in power. And so, I think the committee did good bipartisan work there. And, of course, we reauthorized, for the first time in about a decade, the modernized Safe Drinking Water Act to deal with some of these issues.

And we tackled some of the pipeline siting issues as well. And small-scale hydro and irrigation districts that have put their open canals into pipes, pressurized the systems, and put a little hydro facility in and now generate enough power for 3,000 homes just in central Oregon. So we streamlined some of the licensing there for hydro, which is an area where we get, you know, carbon-free renewable energy. And to your point, we manage that precious water very carefully.

Dr. Ekwurzel, I am curious. You mentioned wildfires. My district is subject to it. As I pointed out, this is habitat. The committee twice held hearings on the human effects of the wildfire smoke. And scientists told us between 2,500 and 25,000 people die prematurely every year from consuming wildfire smoke.

And we had other forest scientists tell us that part of the problem in the west is overstocked stands, that historically you would have 70 trees per acre and today you have 1,000 trees per acre. And, of course, we know trees are pumps, they take water out of the ground.

As you look at some of this science is that—knowing the effects of wildfires—is that something your organization would advocate for, is modern forest management practices to reduce excess fuel loads?

Dr. EKWURZEL. I had the opportunity to be in Oregon with Forest Service scientists while fires were going. And seeing the sort of native practices to maintain more healthy forest reserves, definitely prescribed burns, other types of actors, are really important. At the same time you want to keep the carbon of the forests being a net storage for a long time—

Mr. WALDEN. Right.

Ms. EKWURZEL [continuing]. Rather than we really do need advances in understanding how to keep wildfires safe and keep populations down-smoke, shall we say. Because there were studies that it is almost like smoking several packs of cigarettes——

Mr. WALDEN. Oh, it is awful. Awful.

Ms. EKWURZEL [continuing]. If you are in a summer situation breathing this smoke.

Mr. WALDEN. Yes.

Dr. EKWURZEL. Which we did breathe some of that Oregon smoke.

Mr. WALDEN. We were suffering under this for 6 weeks. Worst air quality in the world, absent Beijing. Or I mean, there were a couple of countries around the world that just at different periods had worse. But my district faced this all summer, summer after summer.

And we know the prescription is going to reduce—we are always going to have fire, we are always going to have hurricanes, what do we do, though, to minimize the impacts? So thank you for that.

Mr. Powell, as you have indicated, we have been pursuing policies on the committee to promote a range of clean technologies from nuclear energy, hydropower, grid modernization, energy efficiency, and battery storage. But, clearly, we all know what work needs to be done.

The chart on page 2 of your written testimony shows the transition to a zero-emissions energy system is not yet happening globally, that clean energy is just keeping up with energy demand. And we heard that, I think, from Mr. Worthington, too, about the demand out there. But nations still strive for simply having electricity.

How do we build on what we have done domestically so far to increase the pace and scale of technological innovation? And can we do this without imposing economically harmful regulations? And how does deregulatory policy help in innovation?

Mr. POWELL. If we are taking a global lens on this problem— first, thank you for your leadership in the last Congress to expand many of these policies—we are taking a global lens on this problem, the key is making clean technology cheaper, not traditional energy more expensive. If we are making clean technology cheaper, then we are focused on things like, to Chairman Tonko's point, moonshot programs to set very aggressive technology goals, for example, at the Department of Energy, and develop most of our resources toward achieving those very aggressive cost and performance goals. And then we can do more to set targeted incentives that work with markets to help scale up these technologies and get some of the scale and learning-by-doing benefits that Mr. Duke discussed.

Then we can still do a great deal, for example, in streamlining permitting for new hydro projects. It still, despite the great work of this committee, takes far too long to put a new pumped hydrostorage facility in place or to relicense an existing dam, or to power up a nonpowered hydro facility.

Mr. WALDEN. It seems to me we have led in energy development, clean energy around the globe. And certainly with fracking and natural gas replacing 16 gigawatts of coal, that has made a difference around the world and here at home. And I just want to see America lead in these efforts. And obviously we know industries are going to have to step up to the plate here too, but I sense they are willing to.

So thank you, Mr. Chairman, again. My time has expired. And I appreciate all the testimony of our witnesses. Thank you for participating.

Mr. TONKO. Thank you. The gentleman yield back.

The Chair now recognizes Representative Peters from California.

Mr. PETERS. Thank you, Mr. Chairman. Thank you for having this hearing.

We all know the causes of climate change. I respect and appreciate hearing from the witnesses. Now we need to identify the practical ways to stop it, whether that is through regulation, deregulation as in the example of hydropower, putting a price on carbon— I think that is probably useful—carbon capture, R&D, or some combination. Some of these are more feasible than others. But let me be clear, feasible is not a euphemism for lack of ambition, it is just the opposite. Feasible means achievable.

And I want to say from the bottom of my core is that we have to do this in a bipartisan way. What I have learned here is that if it is not bipartisan, it won't pass. And if it is not bipartisan, it won't last. And I really want to make sure that we get everyone on board.

If it was up to me, we would enact a national version of SP100, which commits California to 100 percent carbon neutrality by 2045. We would take those steps. It is not up to me. It is not up to any single one of us to do that. So I am looking forward to working with everyone on this committee to make progress.

We know we have to transition to a clean energy economy. There is not widespread agreement in either party what clean energy means. Maybe it's 100 percent renewables to some people, renewable electricity for some other people. And whether renewable electricity is all zero- and low-carbon sources of renewables or net zero, we can talk about that. But there is a need to move.

And I also just want to, finally, note the presence of Reverend Woodberry here. There is a moral component to this too. And I am aware of Pope Francis speaking out on this as well as the Evangelical Environmental Network.

Let me ask a couple questions of the witnesses. I will start with Mr. Powell.

Climate models show that we are going to need significant deployment of current and new clean energy technologies, including renewables, nuclear, carbon capture renewal, removal. While regulation is an important driver for technology deployment in the U.S. to help global emissions reductions, one of the most important things we can do is to lead on clean energy innovation.

What is the Federal Government not doing right now that we should be doing to accelerate the deployment of these technologies?

Mr. POWELL. Well, first let me thank you, Representative Peters, for your leadership, especially in nuclear innovation and cosponsoring the Nuclear Energy Innovation Capabilities Act, which we were pleased to see passed through Congress last year. That set a good precedent for creating a test bed in the Federal Government for developing and expanding these technologies.

And so now I think the next step is, well, how can we go further? And how can we use other powers of the Federal Government to ramp these up more quickly? I think a good idea would be something like the Nuclear Energy Leadership Act, which takes the next step. It sets an aggressive goal to demonstrate multiple advanced reactor technologies within the next decade.

It expands the power of the Federal Government to use its PPA authority to purchase some of the power from those reactors, to get them set up, and to get them financed.

It expands the availability of fuel that they would use.

And I think we could take those kinds of approaches and apply it across all of the different clean energy technologies in order to scale them up more quickly.

Mr. PETERS. OK. I am interested in talking to all of you about deployment as well on other technologies.

Mr. Williams, I believe action on climate change is an opportunity to create economic growth. But it is undeniable that a shift away from fossil fuels will have an impact that is tough on certain sectors. I think we need to provide workers in those sectors with a path to jobs that pay just as well or better, including retirement benefits and protections, the kind of jobs that can support families.

In your testimony you talked about specific things the committee could do in an infrastructure package. What do you see as the most important things for Congress to include in any climate legislation to protect workers?

Mr. WILLIAMS. Thank you for that question, Mr. Peters. We agree completely. That is a critical issue. In my verbal testimony I made sure to lean into the statement that we cannot let any workers or communities be left behind in this effort.

There are a number of ways to do that. And the best way— among the best ways—is to direct the investments that would come from this to workers and communities that may be harmed, but just generally a commitment that we want to actually retain as many jobs as possible, first and foremost. And then, if that is unavoidable, make sure that there is that deep commitment, as you mentioned, to ensure that wages, benefits, healthcare, so on and so forth, people are taken care of throughout that process and that there is significant economic development in communities that see that dislocation.

Mr. PETERS. We have seen, I think, a lot of progress in California that we can learn from as well on that front.

Finally, I just want to say with respect to Mr. Worthington, I haven't had a chance to ask you a question, but we talk about all the people who are underserved in terms of energy around the world, it strikes me that the cell phone is a good thing to look at. You know, a lot of places without phones didn't build out whole set of sort of telephone grids, analogous to the energy grid, they did essentially microgrids with cell phones.

And I would suggest that a large part of our foreign policy should be the deployment and promotion of microgrids, just like the United States Marine Corps has at Camp Pendleton near my district, that don't rely on a centralized fossil fuel-based source but can rely heavily on renewables and on storage. And I think it is very feasible that we should really make that part of the mix.

Mr. Chairman, I yield back.

Mr. TONKO. The gentleman yields back.

The Chair recognizes Representative McMorris Rodgers.

Mrs. RODGERS. Thank you, Mr. Chairman, and all of the witnesses that are here today. I appreciate you being here and sharing your perspective on the environment.

As you may know, I come from Washington State. And we are a leader in hydropower production. And because of research and innovation, new technologies, we are seeing even better salmon returns because of the fish, new, improved fish ladders and turbines. You know, we could double that hydropower without building a new dam in America simply by investing in hydroelectricity also. Only 3 percent of the dams actually produce electricity. And this is a clean, renewable, reliable, affordable source of electricity.

So I wanted to start with a question to Mr. Powell. In the last Congress, I led legislation to streamline the hydropower licensing process. It takes on average 10 years to relicense a dam right now in America, compared to 18 months for natural gas. In your view, how does hydropower fit into the bigger picture? And what are we risking with proposals such as the Green New Deal that ignore the positive environmental benefits of hydropower?

Mr. POWELL. First, thank you, Representative McMorris Rodgers for your leadership on hydropower and preserving and expanding this very important resource. As you know, historically hydropower has been the most important of our renewable resources in the United States, and is appropriately viewed as a renewable energy resource right alongside wind, and solar, and biomass, and geothermal, and other renewables resources.

In many ways it is the most valuable renewable resource for three reasons:

First, it has the highest capacity factor of the renewable resources, so it is available for more of the year.

Second, it is a flexible resource. It can be turned on and off, and ramped up and down in a way that many other renewables resources cannot be.

And third, it can also be part of a storage solution. So pumped hydropower can serve as a, you know, vast battery. In fact, the very largest storage facilities in the United States are pumped-storage hydro facilities.

So we see expansion of hydropower, either by powering up nonpowered dams or certainly ensuring that our existing hydropower facilities around the country are relicensed, and that we can continue to get good use out of them, and modernizing those facilities as key priorities of the clean energy portfolio.

Mrs. RODGERS. What do you think Congress could do to expand hydropower production in the United States? And why do you think that should be a part or a central part of a climate-focused policy?

Mr. POWELL. So it needs to be a central part of a climate-focused policy. As Chairman Tonko said, at this point the climate challenge is too urgent to leave any of our tools off the table. And so certainly the largest renewable resource can't be left out of that solution.

The idea that we would depower all of that hydropower, which I believe powers between 6 and 8 percent of our power grid right now, and replace it with new power, you know, the billions of wasted dollars that would be spent in doing something like that would be very counterproductive to a climate solution, and would certainly not be a cost-effective way to advance climate policy.

Mrs. RODGERS. As we add more intermittent renewables to the grid like wind and solar, grid-scale energy storage will be critical to ensuring a flexible and resilient system that can deliver affordable and reliable electricity to consumers when the wind isn't blowing or the sun isn't shining. I share ClearPath's goals to expand energy storage.

Last Congress, we passed legislation. Mr. Griffith led it. We have also upped research dollars for new, innovative energy technology. I rep—or I am very close to the Pacific Northwest Laboratory. They are doing a great work in this space.

Can you just help us understand more about what is going on in the private sector and what specifically we need to do here in Congress to accelerate innovation in energy storage?

Mr. POWELL. Sure. Well, first I should acknowledge PNNL's leading role in the energy storage innovation space. They have pioneered some of the most promising new technologies that are already being scaled up and commercialized in grid scale energy storage.

I think the first thing to remember is that energy storage is far more than just batteries, right? It can also include things like pumped-storage hydro. It can include innovative ways of using water pressure to store energy underground. It can include heat storage and many other solutions. So I think, first and foremost as we fund against that priority for our Federal R&D engine, we should be thinking of what we want to come out of a storage solution as opposed to the necessary technology that would go into the storage solution.

And I think we can set very aggressive goals against that, as some legislation introduced in the past Congress did, and then drive most of our dollars and coordinated activity across the Department of Energy toward achieving those performance milestones.

Mrs. RODGERS. Thank you. With that, I am going to yield my time. And I appreciate your sharing that info.

Mr. POWELL. Thank you.

Mr. TONKO. The chairwoman yields back.

The Chair now recognizes Representative Barragán.

Ms. BARRAGÁN. Thank you, Mr. Chairman.

Last night at the State of the Union, the President may have ignored the threat of climate change. But with Dems in control of the House, this committee and Congress will no longer ignore the threat of climate change.

I also want to take a moment to thank Reverend Woodberry and Mr. Williams for mentioning the impact to communities of color and low-income communities that climate change is having. When I think of climate change, I don't think in terms of green. I think in terms of black and brown. When I think of climate change, I think of my black and brown constituents who make up 88 percent of my district and who are disproportionately impacted by negative impacts of climate change.

I think of black and brown communities throughout the Nation forced to live under discriminatory environmental policies that cripple their cities and towns economically, and leave them vulnerable and dependant on the very companies that are polluting our neighborhoods.

When I think of climate change, I think of black and brown people who are confined to communities where decades of lax environmental policies and enforcement have literally sickened entire generations. I think of black and brown people across the country, this Nation, who face the painful reality of shortened lifespans filled with health complications caused by the toxic environment in which we live.

I think of black and brown children forced to live in neighborhoods where the air quality standards are astonishingly low and the use of asthma inhalers is alarmingly high. I think of black and brown communities and children whose asthma diagnosis amounts to nothing more than a death sentence, with brown children in these communities having 40 percent or more likely to die from the affliction than their white counterparts.

So, ultimately, when I think of climate change, I do not see an environmental crisis, I see a systematic environmental racism that needs to be acknowledged and addressed.

Reverend Woodberry, do you acknowledge that environmental racism is a real threat to black and brown communities?

Reverend WOODBERRY. Yes. Thank you for your question. Absolutely. And we want to urge Congress that, as we move forward with legislation, we ensure that we are not replicating models of injustice. Let me give you an example.

Last year in August, we cut a ribbon on a solar farm, small solar farm in Dillon County on Highway 9 in the middle of a soybean field. But we were very careful while working with Duke Progress Energy, the utility, over a 2-year period to make sure that this solar farm was built in a just and equitable manner. And so, out of the 1,200 households that will be supplied with energy from this community solar farm, we made sure that one-third of the residents who were 200 percent of the Federal poverty limit had the $250 emission connection fee waived.

And in addition to that, we have to be careful that, as we move toward renewable energy or we do energy grid upgrades, that we are not once again replicating models of injustice. So we were able to get the utility to do 1,500 free energy efficiency upgrades. Because whether an environmental justice home is connected to fossil fuels or renewable energy, if that home is energy inefficient and they are heating and cooling the outdoors and paying a disproportionate amount of their income on energy costs, we have not solved the problem.

And what we want to avoid is creating an energy divide the way that we have done in the past by creating an educational and digital divide.

Ms. BARRAGÁN. Thank you, Reverend Woodberry.

If I could with the last 20 seconds, Mr. Williams, what are your recommendations to the committee to address environmental inequalities in black and brown and low-income communities, including opportunities to create these clean jobs?

Mr. WILLIAMS. Sure. Well, first, thank you so much for your statement and your question. If we put forward a wholehearted effort to solve climate change but in the process do not remove toxic chemicals and other forms of pollution from workers' communities, then we haven't succeeded. So we agree.

So there needs to be a significant, comprehensive effort that incorporates that into efforts to reduce greenhouse gas emissions as well.

In terms of job creation in those communities, absolutely, targeted investments in disadvantaged communities, previously overlooked communities, absolutely needed. Policy items like community benefits agreements, local hire provisions, all are absolutely critical as we invest in trying to find new solutions.

Ms. BARRAGÁN. Thank you. I yield back.

Mr. TONKO. The gentlewoman yields back.

The Chair now recognizes Representative McKinley.

Mr. MCKINLEY. Thank you, Mr. Chairman.

Mr. Powell, I would like to have a conversation with you or some interaction with my remarks here. I think we have heard on the panel so far most Republicans and Democrats agree that there is a—the climate is changing, and that industrial activity is a major contributor to that. But I think the reinforcement is that we strongly disagree with solutions on how that might be.

Would you agree that America acting alone is going to make a difference to the global environment?

Mr. POWELL. It will not.

Mr. MCKINLEY. Thank you. Let me add to that.

So I want to add that, if anyone thinks that decarbonizing America is going to save the planet, whether that is 10 years or 20 years from now, you are delusional. Just 3 years ago, the EPA Administrator said that, her quote was, "American action alone will not make the difference needed to impact global climate change."

The Cato Institute came out and said that decarbonizing the United States would lower the global temperature by just one-tenth of 1 degree Celsius by the year 2050.

But without this global commitment that everyone seems to be ignoring, this is what we are having to deal with. Do we really think, any of you on this panel, that if we decarbonize America we won't be faced with severe weather, we won't have droughts, that coastal communities won't be flooded? How can we say that without the rest of the world on board?

Here is what is going on, as CRS has already published. [Slide follows:]

CO_2 Emissions, 2000-2016

China	+290%
India	+235%
USA	-16%

Source: Congressional Research Service.
Data Source: International Energy Agency, CO_2 Emissions from Fuel Combustion (2018 Edition).

Mr. MCKINLEY. This is what is going on, that China from 2000 to 2016—China has increased its global emissions—or its emissions—by 290 percent. India, 235 percent. And at the same time America has reduced it by 16 percent.

Are you familiar with the MIT report, their technology review report that—maybe you are. And what that said was—and it was just a recent report—it came out and said that, unless India reduces its emissions, the result will be a climate catastrophe regardless of anything the United States does.

I want to make sure we always keep this in context. We don't live in a vacuum. We don't live in a little microcosm here that the air of the United States is, if we can get it clean we will be fine. We involve from the globe on this.

So we get down to, what are our solutions or what are our options? And so if I could from you—you and I have had this conversation—it appears that most of the Democrats or people on the other side of the aisle are saying that they want to use a hammer approach. Let's put more regulations, cap in trade, carbon taxes, some kind of hammer approach. Isn't that what you are hearing as well primarily, Mr. Powell, that it is a hammer approach to solve this problem rather than a carrot and incentives for innovation?

Because I think if we could do the innovation that we started last year with 45Q, with 48A, we could go on with that. Look, we have already talked about the Allam cycle, the net power plant, the turbine efficiency. Aren't those things going to be really the best solution rather than the hammer approach?

Because I am assuming you are aware of the hammer approach throughout Europe, France particularly lately with the yellow vests, what happened there when they rejected that notion of a hammer approach. So, if we could just continue this innovation, this effort for research, I think many of you talked about the research concept, if we could do that we could, America, use our science and technology that we have used to do space, medicine, healthcare, all, and implement a strategy. Wouldn't it be something that we then could export to the other nations so that—like Mr. Worthington was saying, a billion to a billion five that don't have energy—if we develop the technology to reduce emissions and we could see that, export that technology and give them a chance for a better life, wouldn't that make more sense than a hammer approach that people are rejecting?

Mr. POWELL. So technology is the genie you can't put back in the bottle. And the political will for climate solutions will come and go here in the United States and around the rest of the world, but technology will last.

Mr. MCKINLEY. OK.

Mr. POWELL. So we can export the technology and we can have a higher confidence that that will be taken up around the world.

Mr. MCKINLEY. I just hope that everyone on the panel will recognize that what we do here is, we are just part of a big system. We have got to get the rest of the world engaged in this, otherwise we are still going to have severe weather, we are still going to have drought, and we are going to have flooding of our coastal communities.

Thank you very much.

Mr. TONKO. The gentleman yields back.

The Chair recognizes Representative McEachin for 5 minutes.

Mr. MCEACHIN. Thank you, Mr. Chairman. Mr. Chairman, I want to start by thanking you for your leadership in the fight to stop climate change. I can't think of a more important discussion with which to begin the new Congress. And I also would like to thank our panelists, especially Reverend Woodberry, who has been a great champion for environmental justice, and Mr. Williams, whose organization has helped show that organized labor and the environment movement share the same goals and can succeed by working together.

And in that vein, Mr. Williams, I would start with you and build a little bit on the question that Mr. Pallone stole from me, quite frankly.

You know, one of my proudest accomplishments as a State legislator was to help clear the way for an offshore wind farm, which means well-paying jobs for Virginia workers. And I believe that we can replicate that success across the country. So how do we ensure that the coming green energy revolution helps all workers, even those who right now are working in the fossil fuel industry? That is the part I want you to build onto your answer that you gave Mr. Pallone.

Mr. WILLIAMS. Sure. Offshore wind—well, first, thank you for your leadership, Mr. McEachin, it has been extraordinary. And we are already seeing benefits in Virginia for offshore wind investment and those policies working. Offshore wind is an extraordinary opportunity and one where we have seen, especially from the labor movement and the environmental movement, really the cobenefits percolating up in such a beautiful way.

There is only one project currently built. But there are thousands of megawatts on the cusp of being built up and down the east coast. That is going to create high-quality union jobs in coastal areas up and down the east coast. But then going into the country, the supply chain potential of that and helping build out and support American manufacturing is just critical and incredibly impressive.

We think that there needs to be significant support to make sure that that industry keeps moving forward and that policies deployed ensure that these projects are using project labor agreements, that they are, if needed and if possible, targeting it to communities that certainly need economic investment.

So I just couldn't agree more, offshore wind is an absolutely critical part of this conversation.

Mr. MCEACHIN. Thank you.

Reverend Woodberry, we know that environmental injustice hurts minority, rural, and low-income communities. But does facing unique challenges mean those communities also enjoy unique opportunities? For example, if we use the policy process to create new green-collar jobs, can we expect those jobs to be created in an economically just way? And if not, are there steps that we can take to make sure that they are, that they are done in an economically just way?

Reverend WOODBERRY. Absolutely. Thank you for that question.

What we need to do is work on a macro level but also on a micro level so that we are putting in place community-based climate solutions and also doing community in-place training. So we have seen this done successfully in Buffalo, New York, with Push Buffalo where, in the community that was being gentrified, they were able to get a building that was abandoned and convert that building into housing for senior citizens as well as offices for NGOs and a community center.

We also have seen it done, we had some training back in 2017 where we did a train-the-trainer for a solar installation for nonprofit leaders from Georgia, from your State in Virginia, from Mississippi, and South Carolina. And they have gone back in their communities to do solar projects and low-income, people-of-color communities.

As a matter of fact, Monday I had the privilege of speaking at the University of Virginia. And we are going to be launching a solar project in the Buckingham community in June.

And we can actually take these small-scale, community-based successful programs and projects and actually export them overseas. So I cochair an 88-year-old organization called Agricultural Missions, Incorporated. We are just completing an 8-year project in Sierra Leone and Liberia where we brought community water pumps to 47 towns and villages. And we will be going to Sierra Leone and Liberia in April so that we can work with those same community leaders and organizations in these towns that have never had electricity so that we can work on implementing a 4-phase solar project in those towns and villages.

So we can export the technology. We can also export communitybased climate change solutions with renewable energy, providing jobs and opportunities for low-income communities and people of color in this country and around the world.

Thank you for your question.

Mr. MCEACHIN. Thank you, Reverend.

And, Mr. Chairman, I yield back.

Mr. TONKO. The gentleman yields back.

The Chair now recognizes Representative Long for 5 minutes.

Mr. LONG. Thank you, Mr. Chairman. And before I begin my remarks, I would like to ask for everybody to keep John and Debbie Dingell in their thoughts and prayers. Debbie had tweeted out this morning that "Friends and colleagues that know me and know I would be in Washington right now unless something was up. I am home with John and have entered into a new phase. He is my love and we have been a team for nearly 40 years. I will be taking each day as it comes. We thank people for their friendship and support and ask for prayers and privacy during this difficult time."

I know reading this in an open hearing may not be privacy, but she tweeted it so I am assuming that she would be OK with that. And John was sworn into Congress the year I was born, 1955, and Debbie has followed in his footsteps. And very good friends of my wife, Barbara, and I. So just want everyone to keep John and Debbie in their thoughts and prayers, if you will.

I want to focus my questioning here today on how to reduce carbon dioxide emissions while keeping energy and commodity prices low, particularly in rural and agricultural communities like those that I represent. I have a large rural area.

Mr. Worthington, coal represents 81 percent of Missouri's power generation in 2017. And two of the biggest industries in my district are farming and trucking. And from what I have seen with the New Green Deal wants to completely replace fossil fuels with renewable energy and decarbonize our economy, which would be a very worthy goal if it was anywhere near possible within the time frame they want to do it.

Do we currently have any technology to decarbonize the farming and trucking industries while continuing to produce and move goods to market without harming consumers?

Mr. WORTHINGTON. That technology does not exist today at scale to accomplish those goals. We can possibly get there, given time and given tremendous investments in research and technology. Agriculture presents a significant percentage of greenhouse gas emissions. You might think of them as being naturally occurring in the agricultural business. I don't think we are going to change that component over time. There is no technology fix for the emissions out of agriculture.

We have a long, long way to go to develop the technology that would allow for a 100 percent renewable economy.

One recent report that came out in December, part of a scientific journal called Joule, indicated that, if such energy storage options existed, $100 a kilowatt hour for lithium ion batteries, for example—that is a third of the current cost—the cost would be $7 trillion. Seven trillion dollars, just the storage component of a 100 percent renewable system. Seven trillion dollars is 19 times the amount that Americans spend on electricity in 1 year. Nineteen times the amount of electricity in 1 year.

And that would be, again, a cost of lithium ion batteries that is a third of what the cost is now. So, even with additional R&D investments, the cost is still going to be staggering——

Mr. LONG. OK.

Mr. WORTHINGTON. [continuing]. For the Green New Deal.

Mr. LONG. Thank you.

And, Mr. Powell, I will turn to you. And I travel quite extensively with my duties here in Congress. Been to China several times. And I think one time I have seen the sun while I was there. I mean, sun dials are not big sellers because you can't tell if the sun is up or not or what part of the sky that it is in. So anyone in their right mind wants clean air to breathe, clean water to drink, and I hope that hearings like this will bring out commonsense solutions that we can all agree on as Republicans and Democrats and come together to eventually reach these goals.

And, Mr. Powell, I share your desire to reduce carbon emissions, as any right-thinking person would, I would think. And in your opinion what is the right way to do that? Should Congress encourage market-based solutions to encourage cleaner energy? Or should we follow the New Green Deal, which would raise taxes and impose the stringent mandates that have potential costs we just heard about to communities and industries like those that my district deals with?

Mr. POWELL. Well, first, Representative Long, thank you for your leadership on advancing nuclear power and solutions to the spent fuel issue and your work with Leader Shimkus on that issue.

Market-based solutions, all things being equal, should be the more cost-effective solution to the problem both here in the United States and also the things that we can export to other economies like China. It is very difficult for us to export our policy over there. They do their own thing. But they are happy to buy, and take, and scale up our technology. In fact, the real risk is that the Chinese in many of these things are actually moving very quickly and attempting to take also parts of the global market in those technologies as well.

And So I think from the U.S. economic competitiveness perspective, there is a real priority that we stay competitive with these technologies alongside the Chinese.

Mr. LONG. OK, thank you. I am past my time. I yield back.

Mr. TONKO. The gentleman yields back. The Chair now recognizes Representative Blunt Rochester for 5 minutes.

Ms. BLUNT ROCHESTER. Thank you, Mr. Chairman. First I want to thank you for your leadership and also for your charge to the committee that we rise to the challenge. I want to thank you for that. I would like to thank the witnesses as well.

I can think of no more pressing topic for us to be addressing than climate change. Actually, as we were sitting here, over my phone a New York Times article came out to say that it is official, 2018 was the fourth warmest year on record. It is happening to us right now.

And in Delaware we are the lowest-lying State in the country. We are urban, we are rural, we are suburban, and we are also coastal. So the consequences of climate change and sea level specifically impact my State directly.

I also wanted to just say a word about the global conversation that we are having as well. I actually did live in China, and I actually do think that we need to stay competitive. But the real issue is not whether the world recognizes it, it is do we recognize it? When we get out of the Paris Climate Accord, we send a message to the world.

My first question is to Dr. Ekwurzel. And if you can just talk a little bit about the potential impact of sea-level rise for a State like mine if we don't immediately take steps to address carbon emission and climate change more broadly?

Dr. EKWURZEL. Delay in action on reducing global emissions is absolutely critical for the State of Delaware. As you know, the lowlying communities, we also have situations where there are churches that the parking lots—people can't even get to church on Sundays.

Ms. BLUNT ROCHESTER. Yes.

Dr. EKWURZEL. It is really affecting the daily lives. And we have been working with communities to share those stories and to figure out how can we adapt.

Adaptation is really key for the State of Delaware for doing coastal resilience.

Ms. BLUNT ROCHESTER. Thank you so much.

And I would like to turn it to Reverend Woodberry. And thank you also for your work.

One of the things, a lot of people think that sea-level rise really only impacts those coastal communities and beaches. But, as was said, in Delaware we have areas that are considered environmental justice communities. And I was hoping, Reverend Woodberry, if you could just talk about strategies that you have seen that are effective in helping those communities get their voice out there and also advocate for themselves, actual strategies.

Reverend WOODBERRY. Actual strategies. So we have to look at being more proactive rather than waiting for climate impacts to take place. And thank you for lifting that up. And sea-level rise impacts even freshwater. So we are finding waterways, estuaries that are becoming more brackish. It is impacting sea life. It is impacting fishing. A lot of low-income people actually don't fish for sport, but they fish because they need the food in order to survive and feed their families.

Some of the solutions that we discussed recently in New Orleans after experiencing the Hurricane Florence and Hurricane Michael, was that we need to work desperately to put people to work to make our homes more resilient to deal with adaptation. So I mentioned briefly in my statement that we can look at doing bioswales. In a lot of our communities, we have brownfields that are being polluted by industries that are gone that we can actually create bioswales and use plants for remediation that can draw out heavy metals and toxins, and actually provide drainage and pools so that urban areas or rural areas do not have to be as flooded as they are now.

Also, it is very important that we keep our forests and our trees standing, particularly along our river areas. Hardwood is very valuable. But what we are finding is that a lot of low-income communities are actually losing their forests and their trees. We have a lot of folks, particularly people of color, who have their property that is owned by several families, and oftentimes they are not able to pay the property taxes, and the only option that they have is to have the trees cut down.

Ms. BLUNT ROCHESTER. Thank you, Reverend.

Reverend WOODBERRY. So adaptation reserve is really important.

Ms. BLUNT ROCHESTER. Thank you so much.

And, Mr. Williams, my last question is really about, in relation To Reverend Woodberry, many of these communities like Southbridge where we live in Delaware bear the brunt of these economic impacts. Can you talk about jobs that can be created to help mitigate and also strengthen the community?

Mr. WILLIAMS. Sure. And thank you for the question.

Again, this is an infrastructure discussion. This is directing investments directly towards those communities. We should target them to communities that are going to be hardest hit, are already hard hit economically, and we should make sure that we are not just tossing money and saying, "Go forth."

But there should be standards there to make sure that there are good jobs and they are lifting up people who haven't had the opportunities, whether it is building sea walls, or retrofitting buildings, or even working in healthcare and such, just making sure investments get targeted there.

Ms. BLUNT ROCHESTER. Thank you. I am out of time. I yield back.

Mr. TONKO. The gentlewoman yields back. The Chair recognizes Representative Flores.

Mr. FLORES. Thank you, Chairman Tonko. And thank you, Chairman Tonko and Leader Shimkus, for hosting this meeting today. I was pleased that all the panel and almost everybody up here on the dais has agreed that climate change is real. The question is, how do we deal with it?

Reverend Woodberry, I want to thank you for your closing comments where you said that we have got to focus on mitigation, and adaptation, and resilience. And then you further closed by saying that forests are by far the best carbon sink that is available today, and that we need to not forget about that as a source of carbon capture.

I would—I want to say this—you know, we have already heard this, the U.S. leads the world in emissions reduction. And everybody keeps talking about Paris. And the EU countries that are part of the Paris Accord have failed to meet their carbon reductions.

We, on the other hand, have been leaders in this. And it's in large part to technology that has created that American success story, partially because of the transition to cleaner-burning natural gas and the development of cost-effective renewables.

For my own part, I am doing my part. Right before I ran for Congress, I didn't know I was going to run for Congress, but I commissioned the largest residential solar system on my house in Central Texas. And so I am glad to be part of that. And over the course of the last 3 years, I have converted over 90 percent of my light fixtures to computer-controlled LED technology. So I have one of the lowest emissions footprints per square foot of anybody up here on this dais.

That said, you have got to be careful how you do this. I don't think we get it through a chaotic, headlong rush toward decarbonization. I think we get it through thoughtful use of technology and figuring out what is the pathway for this moonshot, and what is the realistic time period that it gets there.

One of the things that—one of the technologies I think that gets us there is nuclear. We hear a lot of projections about replacing the existing fossil energy power generation with solar and wind. But there are mixed messages about the role of nuclear energy in the future. And it seems to me that, if we are really serious about climate change, we need to get serious about the role of nuclear power.

I don't understand why some advocates for that chaotic decarbonization do not take nuclear seriously. They are ignoring the role of next-generation nuclear power as a significant source of baseload zero-emissions power with a much smaller land and environmental footprint than nonbaseload power sources like wind and solar.

Mr. Powell, your organization, ClearPath, is doing a significant amount of work in the nuclear area. What is your organization focused on in this form of clean energy over others?

Mr. POWELL. Well, first, Representative Flores, thank you for your leadership on advanced nuclear energy, both in promoting solutions for advanced nuclear fuel——

Mr. FLORES. We are going to bring it up again, too.

Mr. POWELL. Appreciate that. And also for cosponsoring the nuclear moonshot approach that Representative Higgins has brought to the House Science Committee.

We think that a number of priorities are necessary to scale up the next generation of nuclear power. Obviously, we need the fuel for those reactors.

Mr. FLORES. Right.

Mr. POWELL. We already have a test bed that has now been established in the last Congress. Now we need a moonshot goal to demonstrate multiple advanced reactors and deploy most of our resources through the Department of Energy towards achieving that goal.

We also need to use the full resources of the Federal Government, like its PPA authority to scale it up.

And then, lastly, to this global problem we need to be thinking about how we use nuclear as a tool of diplomacy and economic development around the world and how we use new authorities like the BUILD Act and the Development Finance Corporation to start exporting that good U.S. nuclear technology to other countries and help them solve their emission problems with 24/7/365 clean energy.

Mr. FLORES. The United States is developing advanced next-generation nuclear technologies. But it has also been demonstrated that we have a great record for our current light water reactor fleet. The United States nuclear reactors have operated for over 4,000 reactor years without a major accident, according to the Nuclear Regulatory Commission.

If this knowledge and successful safety record can be shared with the rest of the world, we could make great strides in reducing emissions through safe nuclear power, particularly next-generation nuclear power, to generate clean, zero-emissions electric power.

So, Mr. Worthington—and then I will ask you the same thing, Mr. Powell—should the U.S. promote more nuclear as part of a global emissions reduction scheme?

Mr. WORTHINGTON. Absolutely.

Mr. FLORES. OK. Mr. Powell? Pretty simple answer.

Mr. POWELL. Yes, absolutely.

Mr. FLORES. OK. For both of you, has anyone looked at the environmental impacts of scaling up to 100 percent renewables? My home State of Texas is the Nation's leader in wind production. But then we have got a lot of land, open land in West Texas that makes it feasible to do that where it is not a problem.

Wind, however, is intermittent and does not provide always long baseload power. And so, when we saw that with the impact of the power demands coming out of the recent polar vortex, what are the environmental and land use impacts of wind and solar versus nuclear and natural gas? Mr. Powell?

Mr. POWELL. Well, certainly nuclear is a more compact solution.

Mr. FLORES. Right.

Mr. POWELL. It produces more power on a smaller amount of land. And in terms of the broader environmental impacts, there are tradeoffs with all of these technologies.

So renewable technologies and the storage that would have to go along with them have a lithium problem and sort of a lithium sourcing problem for where they come from. Just as nuclear has a spent fuel problem.

Mr. FLORES. Right.

Mr. POWELL. All of these technologies have their own local environmental impacts, and all of those need to be managed as part of a holistic solution.

Mr. FLORES. Mr. Worthington?

Mr. WORTHINGTON. What we are worried about is, with the rapid deployment of solar photovoltaics, these systems have a shelf life. And after they no longer function, they are going to have to be recycled. And there are some pretty nasty chemicals that are contained when they are manufactured.

And so we are concerned that we don't really have the rules in place necessarily to safeguard that those units are recycled properly and the chemicals are properly disposed of. I think that is something that has not been adequately studied and——

Mr. FLORES. Right.

Mr. WORTHINGTON [continuing]. Warrants some more review.

Mr. FLORES. Thank you. I do agree that, as we have future hearings on this subject, we need to consider the gnarly environmental footprint that some storage technologies have.

Mr. Chairman, thank you for your forbearance. I yield back.

Mr. TONKO. The gentleman yields back.

The Chair recognizes Representative DeGette.

Ms. DEGETTE. Thank you very much, Mr. Chairman.

You know, I really want to thank you for having this hearing as your first hearing of this committee. I have been on this committee for many, many years and seen the, shall I say, evolution of thinking about climate change. And this panel is the perfect example of that.

And so I want to start out in the grand tradition of our beloved friend and mentor, John Dingell, and ask you all a couple of questions that will only require a yes-or-no answer.

The first question is, do you all agree that climate change is real and that human activity contributes to it? Doctor?

Dr. EKWURZEL. Yes.

Mr. POWELL. Yes.

Mr. DUKE. Yes.

Reverend WOODBERRY. Yes.

Mr. WORTHINGTON. Yes.

Mr. WILLIAMS. Yes.

Ms. DEGETTE. Thank you. That in itself is a revolutionary step for this committee. Thank you all for that.

My second question is, do you all agree that we need to address climate change in a way that builds the resilience of our communities, especially of those most vulnerable to climate impacts, while growing our economy and providing well-paying jobs? Doctor?

Dr. EKWURZEL. Resounding yes.

Mr. POWELL. Yes.

Mr. DUKE. Yes.

Reverend WOODBERRY. Absolutely yes.

Mr. WORTHINGTON. Yes.

Mr. WILLIAMS. Unequivocally.

Ms. DEGETTE. Thank you. My last yes-or-no question—so far you are all getting 100 percent. My last question is, do you agree that driving innovation in clean energy is an essential part of the solution, and that it is time that we committed ourselves to doing that?

Dr. EKWURZEL. Yes.

Mr. POWELL. Yes.

Mr. DUKE. Yes.

Reverend WOODBERRY. Yes.

Mr. WORTHINGTON. Yes.

Mr. WILLIAMS. Yes.

Ms. DEGETTE. Thank you very much.

You know, all of this agreement here in this panel with the Democratic and Republican witnesses makes me really hopeful that, as what Mr. Powell said, bipartisan cooperation on climate change can be attainable. And I want to thank all of you for committing to this.

I just have a couple of more questions.

One of my questions for you, Dr. Ekwurzel, is, as you know, I am from Colorado and the last few years we have had the 30-year low in snow pack. And what is even worse than that is that the snow is melting earlier, and so the water is going down. Can you let us know what kind of impact climate change is going to continue to have on the snow pack in the western United States?

Dr. EKWURZEL. Thank you. And that snow pack is a critical water resource for Coloradans and all downstream——

Ms. DEGETTE. Right.

Ms. EKWURZEL [continuing]. In the Southwest.

I want to say that there are three things that climate change does to the snow pack. It causes it to melt earlier. We have a shorter snow season. Even if you have an atmospheric river delivering wonderful amount of snow, the extra heat in the winter season is causing it to melt, and sublimate, and evaporate into the atmosphere.

We have what is called a hot drought in the Colorado River. We could lose up to 50 percent of that flow just from the climate change impacts if we were to do unabated, you know, course that we are on now.

Ms. DEGETTE. Second, so thank you, a second issue that we have, in particular in my congressional district, which is primarily Denver, is a persistent smog problem. And of course we all know what the issues with smog are in terms of asthma and the work and school days, outdoor recreation days, et cetera. But what can you tell us—and you talked, we talked a lot and we know in the West about the impact of wildfires—what can you tell me about the impact of climate change on air pollution and smog?

Dr. EKWURZEL. We call it the climate penalty of smog. One of the ingredients you need for greater ozone ground-level production is warmer temperatures. The warmer it is, the more smog you produce if you have those precursors of volatile organic carbon. And you need sunlight.

Therefore, if we were to reduce global emissions, we would reduce the future climate penalty that could only get worse with climate change.

Ms. DEGETTE. Thank you.

Mr. Chairman, I just want to respond to a couple of the things our colleagues on the other side of the aisle have been saying. The first thing they have been saying is that, well, the rest of the world is not coming along.

Well, number one, we are the ones that pulled out of the Paris Climate Accord, not them. And so I would suggest maybe one of the first things we could do is get back into the Paris Climate Accord.

And the second thing I will say is, just because other people aren't moving as quickly as we are, the President said last night in the State of the Union, America is the best country in the world. Why don't we be the trendsetter? Why don't we be the one exporting all of our technology to China and India? Why don't we be the one setting the standard?

And the last thing I will say is, these other countries do want to act. Their citizens are demanding action for the same reason why we are demanding action. And I think that that is why this committee—this is just the first step, and I know you intend to work on legislation, and all of us intend to work on that with you because we are actually going to move this through. And I know we can do it in a bipartisan way.

I yield back. Thank you, Mr. Chairman.

Mr. TONKO. The gentlewoman yields back. And we thank you for your comments.

The Chair now recognizes Representative Carter.

Mr. CARTER. Well, thank you, Mr. Chairman. And thank all of you for being here. This is an extremely important subject. I believe that my colleague just asked all of you a question on whether you believe that climate change is real or not. And I think, if you were to ask that same question to everyone up on this dais, they would say the same thing: Yes, it is, it is real. It is something that we have to address.

There may be some difference of opinions on how much of it is man-made. But regardless of how much of it is man-made, we still have to address it. There may be some who want to say that it is just cyclical in nature and that if you look back over time and this happens, well, that may be true too. But regardless of that, we still have to see the impact and have to address the impact that man is having on this.

These are all givens. These are all things that I think all of us agree on and all of us are working toward.

I want to start—and for that I want to thank all of you for being here and thank all of you for your interest and for your work on this, because it is extremely important. We all recognize that.

I want to start, if I could, with Mr. Worthington and just ask you, I have always been one who subscribes to an all-of-the-above-type energy policy. I think it is extremely important for a number of reasons for us to have safe, and secure, and dependable, and affordable energy. And it is important for our national defense. It is important for our citizens. It is just very important.

I know that you mentioned in your testimony that you believe that an all-of-the-above approach is essential as well. Once we get beyond solar, wind, hydro, geothermal, and even beyond the traditional fuels, what are some areas that we should be looking for to play a greater role in the all-of-the-above fuels mix?

Mr. WORTHINGTON. Well, one of the, one of the promising technologies is hydrogen. And we have been dealing with hydrogen for decades now. We are not at a stage where it is economical, but it has tremendous potential, both to serve transportation issues as well as electricity. It needs more work. It needs more research. But it is a very promising area that we are watching very carefully.

Mr. CARTER. What about biomass? Let me ask you about that. I represent South Georgia. We have got a number of things in abundance in South Georgia, one of which is pine trees. And we have got a number of biomass manufacturers. And what about biomass, is that something we should be looking at?

Mr. WORTHINGTON. Absolutely. We are actually using biomass now in many different applications. We are using it directly to produce electricity. We are mixing it with coal to reduce the CO_2 emissions from a coal plant. And we are actually pelletizing wood and shipping it to Europe. There are many, many countries in Europe heat their homes with American wood.

Mr. CARTER. Why is that it is used in Europe but not necessarily as much here in America? I always found that interesting. I have visited a number of these plants in South Georgia, and that is what they tell me: We ship it to Europe.

Mr. WORTHINGTON. Yes, it is a very good question. I am not sure I know the answer. It may be a matter of convenience. Our industry has made heating with fuel oil and natural gas very convenient. We have liquefied petroleum gas.

I think it is—I have never answered that question before, but I would have to say it is probably because we have more options than what the Europeans do. And particularly now with our abundant shale gas resources, we are just literally awash in gas.

Mr. CARTER. Right.

Mr. WORTHINGTON. And it is inexpensive, it is affordable, and it is going to be available.

Mr. CARTER. OK. Mr. Powell, I am going to you and ask you, and to kind of follow up on my colleague from Texas, nuclear power is certainly something I feel like we need to be looking at. Georgia Power right now has the only two nuclear reactors under construction in our country. That is something that we are depending on and something I think we should look at very carefully.

Can you tell me the role that you see nuclear power as playing in our country's energy future?

Mr. POWELL. Absolutely. And, first, let me thank you for your leadership in nuclear power, for the State of Georgia's commitment in getting those reactors built. That is incredibly important for keeping the national nuclear supply chain robust and strong going forward.

I think the next generation of nuclear power in the United States will be much smaller, less capital intensive, and more flexible. So I think the future of nuclear power——

Mr. CARTER. We are certainly glad to hear that in Georgia.

Mr. POWELL. Yes, exactly. I think it is unlikely we will build more gigawatt-scale reactors like the great technology going up in Plant Vogtle. I think it is much more likely we will build small modular and microreactors that can be combined together in the same way that wind turbines are combined together in large arrays with hundreds of units. I think that is the future of nuclear power.

Mr. CARTER. Right. Again, let me thank each of you for being here. I appreciate it. This is extremely important, something that we all agree on that we have to address in a reasonable and a rational way that is going to provide for safe, secure, dependable, affordable energy for our citizens.

And I yield back, Mr. Chairman.

Mr. TONKO. The gentleman yields back.

The Chair recognizes Representative Schakowsky for 5 minutes.

Ms. SCHAKOWSKY. Thank you so much, Mr. Chairman, after 6 long years, having a hearing directly on global warming, on climate change. And I wish it were that all of us agreed. Maybe this tweet from the President, who never mentioned this crisis last night in the State of the Union, is a joke. I would like to think so, but maybe not.

During the polar vortex he tweeted, "What the hell is going on with global warming? Please come back fast, we need you!" Not so funny to me. I was in Chicago at the time anyway.

But I want to talk about transportation and its contribution to climate change. The transportation sector is the largest source of carbon pollution in the United States, and only getting worse. And I am very interested in improving our fuel economy standards and decreasing carbon emissions.

The past four decades the corporate average fuel economy, what we call the CAFE standards, have been an extremely valuable tool in reducing greenhouse emissions. Unfortunately, this administration is attempting to weaken vehicle fuel.

So let me ask you, Dr. Ekwur—you know who you are. I will leave it at that. If you could talk to me about the importance of the CAFE standards and making them perhaps even stronger than they are.

Dr. EKWURZEL. Absolutely. We do need to double down on lowering the carbon, decarbonizing our transportation sector, increasing incentives for electrification of the transport sector in cars, and buses, and trucks.

And what we see is that it is also going to lower the ground-level smog as well. It lowers emissions to the atmosphere that causes climate change.

And also, we improve the health of incentives, reduce the inequities with asthma sufferers and so forth.

Ms. SCHAKOWSKY. I am wondering if you can explain this to me. What we have seen over the recent years, some decrease in carbon emissions and global emissions, but we saw last year just in the 1 year that internationally 2.7 percent increase over the previous rates. One scientist called it a speeding freight train. And then in the United States last year, 1 year, marked the largest increase in 8 years, 3.4 percent increase.

So what is going wrong here?

Dr. EKWURZEL. Absolutely. The U.S. was decoupling our growth from a high-carbon economy. We have a lower-carbon economy.

However, that turned around and now the U.S. is emitting more than it did in the prior few years.

So we cannot take our foot off the pedal, so to speak, on incentives that reduce and have cleaner options for when we move around, or power, or turn on the lights.

Ms. SCHAKOWSKY. Thank you.

Mr. Duke, what impact will rolling back efficiency standards have on greenhouse emissions?

Mr. DUKE. Thank you for your attention to the extraordinary benefits that come from fuel economy standards on light-duty and heavy-duty vehicles. And if we just look at the sweep of history on this program, I think it is important to recognize that it was actually Republican President Ford who put in place the first commitment to double our fuel economy back during the initial oil crisis.

And that worked. We got immense consumer benefits and national security benefits out of those efforts. Unfortunately, we then hit the skids on the program when we failed to update the standards for a 25-year period until 2010. And that cost us by some estimates a trillion dollars in additional expenditure at the pump.

So the good news is that we have a set of standards now in place for heavy-duty vehicles that are proceeding and that are going to be helping us transition to advanced technologies for super trucks and the like that will save quite a bit of fuel for industry and our economy.

The bad news is, as you suggested, there is a rollback under consideration which, frankly, goes much further than the automakers themselves requested in engaging with the administration on this. And that is because they know that they need to compete with China. China already has 60 percent market share on electric vehicles. Our automakers need to be competitive, and they can be competitive. Tesla retains the number-one spot. GM is in the top 10. But we need standards that are clear and steadily improving to drive progress and make sure we stay in the game on technology. And fuel economy is part of that.

Ms. SCHAKOWSKY. Thank you. I want to thank all the panelists. This has been really enlightening.

Yield back.

Mr. TONKO. The gentlewoman yields back.

The Chair recognizes Representative Duncan.

Mr. DUNCAN. Thank you, Mr. Chairman.

If the Green New Deal policies are adopted, the price of utilities will inevitably go up. How would the increased cost of utilities as a result of this proposed Green New Deal—carbon tax, cap and trade, high costs associated with renewable energy generation—improve the lives of, say, those in Marion County, South Carolina, that Reverend Woodberry spoke of? People who Reverend Woodberry said were living on fixed incomes of $600 to $800 a month.

The average median income in Marion County is $30,562. And the average median income in my district is $47,000 a year. But the carbon taxes levied on South Carolinians' electricity, gas, et cetera, will increase. These increased costs will impact every person and business in the State and, unfortunately, would disproportionately impact those in the lower-income communities.

And at the end of the day, people care about things that are tangible to them: how much it takes to fill up their gas tank, how much their electric bill will be, and if they have any money left over at the end of the month to put food on the table. That is what my constituents care about.

We here in America, we take for granted what is known as 365/24/7 baseload power supply always on. That always-on power is generated primarily in three ways: hydroelectricity, nuclear power, and fossil-fuel-generated power. Everything else is intermittent. The sun doesn't always shine, the wind doesn't always blow. And we don't have the technology available yet to hold large quantities of power in some sort of battery to provide power when it is needed. We take for granted that 365/24/7 baseload always on power.

But there are people all over the globe that don't take advantage of that. And those are in some European countries, by the way. But think about how the United States can be a leader in improving the quality of lives of so many people around the globe with the export of our fossil fuels so that these folks can have always-on power.

Think about the infant mortality rate across the globe where people don't have a steady 24/7 baseload power supply. They can't keep the incubators on to keep the babies alive.

If we want to improve the quality of life—Mr. Worthington mentioned 1.3 billion people in the world don't have power—think about the quality-of-lives issues that he was bringing up earlier. Air quality. Air quality kills, what, 400,000 people around the globe annually because of bad air quality. They are cooking on charcoal, and dung, and wood products. They can't keep food fresh because they don't have electricity to have a refrigerator to keep the food fresh, so the foods spoil, and they are having to eat it and constantly replenish it.

They can't keep windows in the window spaces because they don't have electricity to provide air conditioning, so at night they are trying to keep cool, mosquitoes fly in. When mosquitoes fly in, they bring diseases that kill so many people around the globe every year.

Food safety, preparation of food, cooking of that food, air conditioning, lights to read to their children and have their children read to them, these are quality-of-life issues that we take for granted here in America that fossil-fuel-generated power can provide for people around the globe. But yet we want to vilify and demonize fossil fuels that make our lives so much better.

Doctor, you are from Massachusetts; right?

Dr. EKWURZEL. I live right here in DC.

Mr. DUNCAN. OK. Well, Cambridge, Massachusetts, is where the organization is located?

Dr. EKWURZEL. Yes.

Mr. DUNCAN. Unless you all rode a bike here today, you came in some fossil-fuel-generated power, whether it was an electric car, probably the electricity that went into that car was provided by some sort of power generation. Could be nuclear, could be hydro, but generally it is probably fossil-fuel-generated.

Many people in this room who came to this hearing today may have gotten on an airplane. And I know just about every Member in this committee got on an airplane to fly here. An airplane is running on a fossil fuel. Folks, your cars, your trains, your planes, are all generated, are all powered by fossil fuels. And we have got a lot of work to do if we are going to make those airplanes fly on electricity. We have got a lot of work to do if we are going to provide electricity through intermittent power supplies to give us that 24/7 baseload power.

But it is not the Government's role to incentivize or penalize companies and individuals that aren't investing in this, it is up to the marketplace. And I am going to use Elon Musk, because I think he is a leader in two areas. He is a leader in EVs with Tesla, but he is also a leader in space exploration. And guess what? He is not being incentivized that I know of for space exploration. He actually said let's pull away from NASA and the bureaucracy and let's think outside the box and figure out how we can save costs, make renewable rockets so that we can travel to the moon and then, ultimately, to Mars. He didn't do that with the Government forcing him to do it. And he didn't do that with the Government incentivizing him to do it. He did it because he had a desire to do that, and he brought the best people together in a capitalist, free market environment to think and come up with a solution.

That is the solution if we truly believe in global warming and improving the lives of so many people around the globe. We do it through the innovation and the innovators, not through punitive or incentives from Government.

Mr. Chairman, with that I yield back.

Mr. TONKO. The gentleman yields back.

Reverend Woodberry, you were made mention of. Do you want to respond in a minute or less, please?

Reverend WOODBERRY. I will say that I do believe that innovation, I do believe that America could move quickly. My family is actually from Marion County. In the 1960s my grandparents, my grandfather was a sharecropper. He used kerosene lamps. They had a stone fireplace and a wood-burning stove for heat. In 20 years we went from having two roads paved to every road paved, everybody moving from outhouses to indoor plumbing. No more kerosene, but instead having electricity for everyone.

We can move quickly and we can use technology. We can use the Government to help because that is who made this happen.

Thank you.

Mr. TONKO. The Chair now recognizes the gentlewoman from California, Representative Matsui.

Ms. MATSUI. Thank you very much, Mr. Chairman. I really do appreciate the witnesses here today.

I find this really refreshing at this point, because I think everybody believes that climate change is real. There seems to be that agreement. And I think that is, in essence, great progress. This is agreement of a National Climate Assessment, which really said that it is real and the risk is now.

And it really concludes that greenhouse gas emissions from human activities are the explanation for global warming over the last 60 years. And for the second year in a row, the transportation sector was the largest emitter of greenhouse gases in the United States. And the International Energy Agency has found it is the only sector that has become less energy efficient over the last 15 years.

My colleague Ms. Schakowsky brought this up, and I want to have a further conversation on this about fuel economy and decreased auto greenhouse emissions. That is what the Obama administration did for light-duty vehicles through 2025, and how important it is in combatting climate change. These standards were written in 2012 with the support of the auto industry, the environmental groups and the States.

Now, these are good for consumers, who save billions of dollars at the pump over the life of their vehicles. And they are good for the American workers, who benefit from the development of innovative technologies that create profits and support jobs. The standards are projected to reduce gas emissions by 540 million metric tons and reduce oil consumption by 1.2 billion barrels, and nearly double the fuel economy of light-duty vehicles to an average of about 54 miles per gallon.

Now, at a time when our country desperately needs to become more resilient when it comes to adapting to climate change, I am really disappointed that the Trump administration moved to reverse much of our progress with their proposal to roll back the curtain on fuel economy and greenhouse gas standards. And that is why I was pleased to introduce the Clean and Efficient Cars Act yesterday which will protect our fuel economy and greenhouse gas emission standards through 2025.

My legislation maintains the Federal Government and auto manufacturers' promise to the American people, a promise for clean, efficient cars that cost less at the pump, better for the environment, the health, and the future of our children and grandchildren.

Mr. Duke, you mentioned in your testimony that, despite our clean technology edge, the United States is not moving quickly enough to reduce carbon pollution. What effects do you believe the Trump administration's proposed rule to freeze the current fuel economy and greenhouse gas standard have on climate-related environmental impacts?

Mr. DUKE. Representative Matsui, thank you for the question and thank you for your leadership on this crucial topic. It is absolutely correct that the transportation sector has now emerged as the most emitting sector of our economy. And it is one where there are extraordinary solutions today and on the horizon to deal with the challenge.

What industry needs in order to scale up these solutions is clarity and certainly against which they can make their investment decisions. And we had that, for example, in that President Ford's initial push to double fuel economy the first time——

Ms. MATSUI. Right.

Mr. DUKE [continuing]. Provided exactly that clarity. And we saw the industry deliver. We saw the Big Three at that time deliver.

Once again we have the potential to double fuel economy with the 2010 standards for light-duty vehicles and, with that, also move into the electric vehicle competition with China in a complete way where I am confident that our automakers can win the day.

What is troubling is that, with the proposed rollbacks—which, again, really exceed what industry itself was calling for, maybe not what certain other industries were calling for but what the autos themselves were calling for—with those rollbacks, it basically makes it harder for us to compete in this global marketplace. Again, China has a 60 percent electric vehicle share, so we don't want to cede that ground.

And I should also note that there is plenty more that can be done and should be done to improve internal combustion engine vehicles as well. There are opportunities to cut emissions from those conventional vehicles much more than we already have today, and cost-effectively. And so we need to stick with the plan that we had in place and keep that investor certainty in place so that we can continue to compete.

Ms. MATSUI. Exactly right. Because we keep moving forward and we have the momentum, and we have to pull back. Business does not like a lack of consistency. We all know that.

Mr. Williams, you mentioned in your testimony that millions of American jobs depend on continuing American leadership on clean vehicle technology that includes over 250,000 Americans employed across 500 U.S. factories and engineering facilities that build technologies that improve fuel economy and reduce pollution. Can you really on a global scale discuss what this will do, just this simple kind of pullback that we have?

Mr. WILLIAMS. Sure. One of the immediate impacts of it, the agency's own analysis says that it will cause, result in the billions less in technology investment that supports 50,000 to 60,000 jobs in the U.S. that we would immediately potentially lose.

But the other piece of it is that this is devaluing the investment that a number of other companies across the supply chain have made based upon those 2010 standards. So, whether you look at ALCOA making aluminum in Iowa and Tennessee, or ArcelorMittal Steel making steel for the auto sector in Illinois, those investments they made because of the need and the standards set forth to make more efficient vehicles. If we step back, countries like China and countries in Europe and throughout the world will take over this industry and completely leave us in the dust.

Ms. MATSUI. Thank you. I have run out of time. I yield back.

Mr. TONKO. The gentlewoman yields back.

The Chair recognizes Representative Johnson, 5 minutes.

Mr. JOHNSON. Thank you. Thank you, Mr. Chairman.

You know, last Congress we began a discussion on our domestic nuclear industry's ability to compete on the world stage, particularly with state-backed enterprises coming from countries like China and Russia. I hope to continue that discussion in this session of Congress.

And I also would like to point out a similar issue occurring on the coal front. As Mr. Powell's testimony states, China is financing about 100 gigawatts of coal projects in at least 27 countries. Like with our nuclear energy deployment, I worry the U.S. is missing an opportunity here, especially as ongoing public/private work is driving down the cost of carbon capture and storage technologies, as well as making nonsupercritical projects feasible here in the U.S.

In other words, the United States is capable of solving these technological problems, but we have got to make sure that we stay engaged on the global front in doing that.

So, Mr. Worthington, can you discuss why so many countries are looking to China for their energy needs?

Mr. WORTHINGTON. Yes, sir. Thank you very much for that question.

The World Bank made a decision a couple years ago that they were going to refuse to consider financing for a new coal plant. There are countries in the world that coal is their only option. Kosovo is a great example. Kosovo has a 50-year-old coal plant that badly, badly needs to be replaced. The World Bank made a commitment to finance a new project.

And as soon as they made that commitment, they started figuring out how they were going to get out of their commitment.

The Chinese have stepped in in Asia, Africa, and South America, and they have been willing to finance projects that the World Bank refuses.

Mr. JOHNSON. And I have heard from our State Department and from our former U.N. ambassador, Ambassador Haley, China is doing this kind of stuff.

Mr. WORTHINGTON. Right.

Mr. JOHNSON. I mean, they are doing this kind of stuff all over the world, all over their region. And they are using these energy projects as a way to get their foot in the door. And then they have big influence in those countries.

So are the technologies supplied by China the most advanced fossil technologies in the world?

Mr. WORTHINGTON. Not what they are selling to other countries.

Mr. JOHNSON. Right. Exactly. Would it benefit these nations if the United States participated in these markets, could we bring the best to the table?

Mr. WORTHINGTON. There is no question. And the other thing that the Chinese do is, they insist that the developing country buy Chinese products.

Mr. JOHNSON. OK.

Mr. WORTHINGTON. So they are not just financing, they are providing all, they insist on providing all of the equipment.

Mr. JOHNSON. Right, right. So how can the U.S. do better from an international engagement standpoint? What should we be doing?

Mr. WORTHINGTON. Well, we have tools ourselves with the Export-Import Bank, Overseas Private Investment Corporation, Trade Development Agency and so forth. Some of these U.S. agencies over the last number of years also adopted an antifossil energy——

Mr. JOHNSON. Right.

Mr. WORTHINGTON [continuing]. Approach. I believe that is being reversed. And I believe that they are open for business now for fossil projects.

But the key becomes the new president of the World Bank. President Trump should identify a new president of the World Bank shortly. Hopefully he or she will not have the antifossil bias that the predecessor did.

Mr. JOHNSON. OK. Mr. Powell, have you got any comments on that before I move on to another question quickly?

Mr. POWELL. I think we can use the new instruments that we created in the BUILD Act, like the Development Finance Corporation. And to your point about sort of China using this strategically, I think we should remember with a nuclear plant, for example, 10 years to build, 80 years to operate, 10 years to decommission. That is a centurylong relationship——

Mr. JOHNSON. Oh yes.

Mr. POWELL [continuing]. That they are getting with that other country. We have that opportunity as well in so may countries, and it does seem like we are squandering that opportunity.

Mr. JOHNSON. Absolutely.

Mr. Worthington, your testimony states that natural gas emissions have declined while production has increased. And that is thanks primarily to technological innovations throughout the industry. I know eastern and southeastern Ohio have benefitted greatly from this increased production, especially as proposed new ethane crackers and other new job opportunities, ethane storage hubs, et cetera, continue to emerge.

So how can we ensure other countries and the world benefit from these technological advances? And what role can U.S. LNG play?

Mr. WORTHINGTON. U.S. LNG can play a pivotal role. We have got a couple units exporting now. We have four more that are coming online either still this year or the early part of next year. We have an opportunity to more than double our LNG exports and to countries like Poland, China, India, Italy, even the U.K. So it is a tremendous opportunity.

We are a dependable supplier. We don't use LNG, we don't use natural gas as a political weapon the way some of our competitors do. And we should just do everything we can to expedite the next fleet of LNG export facilities.

Mr. JOHNSON. Yes. Russia in particular, they get about, Mr. Chairman, they get about 50 percent of their revenue from the sale of oil and gas, much of that to our allies in the region.

I yield back. Thank you very much.

Mr. TONKO. The gentleman yields back.

I now would recognize Representative McNerney from California for 5 minutes.

Mr. MCNERNEY. I want to thank the chairman and I thank the panel for your testimony this morning.

First I would like to observe how reasonable the Republicans sound today on the issue of climate change. There must have been a conversion on the road to Damascus recently.

Dr. Ekwurzel, do you agree that most or all climate models consistently underpredict the climate change rate?

Dr. EKWURZEL. Yes. Because there is a double-edged sword of uncertainty with climate change. The best-case scenario is, we could do that well. But the worst-case scenario tends to keep surprising us. It is a bigger error bar on that.

Mr. MCNERNEY. And given the lag between CO2 emissions and its impact on the climate, do you believe there is a realistic way we can avoid temperature increase of less than 2 degrees C by carbon reduction emissions alone?

Dr. EKWURZEL. We have to have a mix of emissions reductions, all sources of carbon storage as well that we can think that is safe for communities so we can get to a net-zero situation by mid-century.

Mr. MCNERNEY. So then what our alternatives to reduce emissions to avoid climate catastrophe? What are our emission alternatives?

Dr. EKWURZEL. As been said, we have to manage our forests so that they don't go up in flames and lose the carbon they are sequestering. We have to increase the land sink in agriculture practices. We also have to perhaps carbon capture and sequestration, there may be a bridge for innovation through utilization; however, it has to transition. We have to figure out to sequester the carbon and keep it out, away from the atmosphere.

Mr. MCNERNEY. Well, considering climate intervention or geoengineering such as injecting sun-reflecting particles into the stratosphere, how much understanding do we have of climate intervention as to its effectiveness or its possible side effects?

Dr. EKWURZEL. We have a lot to do with the social sciences of the governance of such an issue of just injecting stuff into the stratosphere that would affect perhaps monsoon rains and all sorts of consequences around the world and give us perhaps hazy skies, beautiful sunsets but hazy skies and other consequences. We need more research in this space before.

Mr. MCNERNEY. Well, what do we need to do to develop sufficient expertise in climate intervention to even decide if it is a possible way to manage climate change while we reduce our carbon emissions?

Dr. EKWURZEL. First of all, make sure we invest in NASA and NOAA and our infrastructure to make sure that every time a volcano emits anything that we are able to track it and figure out what the consequences are, because that is the modern, the natural analog to what these experiments would say. And there are many other ways we can study this problem before we would do some other experiments.

Mr. MCNERNEY. Well, then, do agencies such as NOAA and NASA and the DOE have the capabilities to generate a baseline understanding of the stratosphere?

Dr. EKWURZEL. Absolutely. And there are sensors and satellites we would love to have deployed and to double down on science investment on these persnickety problems, as you say.

Mr. MCNERNEY. Well, I might be proposing legislation to do that. And before I finish, I just want to say, Mr. Shimkus, thank you for attributing the quote to me that it is just an engineering problem. But I have to say that was taken out of context. I was referring to nuclear waste being an engineering problem, but I also said that nuclear waste will need a political solution. Now, that whole context also applies to climate change. There are engineering solutions that need to be addressed, but we need to have the political will to put those solutions into effect. And so instead of just sounding reasonable, please work with us to find solutions that are sufficient to the threat.

I yield back.

Mr. TONKO. The gentleman yields back.

The Chair now recognizes Representative Ruiz of California for 5 minutes.

Mr. RUIZ. Thank you, Mr. Chairman.

Dr. Ekwurzel, the National Climate Assessment outlined many severe public health effects of climate change due to increases in air pollution and expansions in the ranges of disease-carrying organisms. I ask this question because I am an emergency physician with a public health expertise as well.

In addition, a study recently published in the New England Journal of Medicine by Haynes and Christie found that in the United States it is estimated that almost 60 percent of the excess deaths may be caused by the use of fossil fuel from power production and traffic. A previous study in 2009 from the same journal, the New England Journal of Medicine, found that a decrease in air pollution is associated with an increase in life expectancy of more than nine months.

This is real. This has real effects for individuals back home when they ask, how does this affect me? It is not an esoteric, ideological, partisan kind of conversation. This is real, pragmatic life effects on your relatives and your children.

In Riverside County, where I am from and represent, ranks amongst the worst in the Nation for ozone pollution. High-ozone days contribute to many hospital admissions, especially for children who suffer from asthma, and seniors with COPD. I know because I personally have treated many of them in the emergency department.

Let me ask you a question. Isn't it true that climate change is making it more difficult to improve air quality?

Dr. EKWURZEL. Yes. The ozone, ground-level ozone with higher temperatures, we call it kind of a climate penalty on health.

The other thing is that Southern California and Arizona have a situation with the extra dust, and the conditions in the spring lead to something that is called a Valley Fever that people can be in hospital emergency rooms. We lose lives to things that are climate influenced.

Mr. RUIZ. And as a public health expert, I am concerned about the impact climate change is having on the spread of vector-borne diseases. Is it true that climate change is expected to influence the spread of vector-borne diseases? And what kind of new illnesses will Americans be at risk for and/or have succumbed to more?

Dr. EKWURZEL. What we see is that a lot of the pests and some of the disease-carrying situations in the tropics are moving into southern parts of the United States.

Mr. RUIZ. Like what?

Dr. EKWURZEL. Such as dengue fever and other mosquito-borne illnesses.

Other things like West Nile Virus that used to be in a part of the U.S. is now spreading northward and westward.

Mr. RUIZ. Yes. So dengue fever, describe the symptoms, would you?

Dr. EKWURZEL. Yes. I defer to your medical expertise on those symptoms.

Mr. RUIZ. Well, I mean it is not pleasant, put it this way. So because we are running out of time.

As a physician I have seen firsthand that the public health infrastructure serving people in rural areas and in other vulnerable communities, underserved communities, is often underresourced and overburdened, working over capacity. And the residents of these areas, like in my district, are often coping with multiple challenges that make their health conditions more severe.

So the National Climate Assessment discusses the special problems and increased vulnerabilities of individuals in underserved communities. Can you describe these problems?

Dr. EKWURZEL. Sure. Climate change exacerbates the historical inequities. And we have to consider these solutions to help. Lowincome communities, children, older adults, people of color are often at greater risk. And low-income communities are often exposed to these risks and due to historical decisions.

And the health impacts, it is really important that we ensure the vulnerabilities of front-line communities are identified and extra precautionary measures are taken to keep people safe.

Mr. RUIZ. So oftentimes decisions are made by, you know, governments or corporations to start a business with some potential air pollution without the consent or the meaningful consultations with the communities that they are going to affect currently and in the long term. These communities, like those in my district, have a very bad physician shortage crisis. They don't have clinics to go to. They already are experiencing high asthma rates because of the living conditions in which they exist. And they face a higher morbidity and mortality at a younger age than other folks.

That is why I introduced an Environmental Justice Act which will specifically address this issue for vulnerable populations with Senator Cory Booker. We have introduced that together.

So we are all well aware that prevention is far less expensive than treatment and is obviously much more beneficial to patients. I hope we will listen to the warnings of the National Climate Assessment and the IPCC report and start to address climate change. It is not only an environmental problem, it is clearly a significant public health threat with real consequences for real people. I know, because I treated them in the emergency department.

Mr. TONKO. The gentleman yields back?

Mr. RUIZ. Yes.

Mr. TONKO. The gentleman yields back.

And we recognize Representative Soto from Florida for 5 minutes.

Mr. SOTO. Thank you, Mr. Chairman. First I want to thank Dr. Ekwurzel for defining the challenge that we have to avoid surpassing 1.5 degrees Celsius. Global carbon dioxide emissions would have to drop around 45 percent below 2010 levels by 2030, and reach net-zero emissions by around 2050.

And, you know, I was thinking about those dates. And it may seem far off for a lot of us. However, I want to put it in perspective. And we have a special guest that I want to recognize here, Lincoln, who just came in. A name that both Democrats and Republicans can get behind, by the way. So, by 2030, Lincoln will probably be just a teenager by then. And by 2050 he will be in his 30s. Relatively young and still starting his life.

This question, this challenge is not about the folks behind the dais. It is not about most of the folks in the audience. It is about Lincoln and his generation and what we are going to do. In 2050 we are going to look back and say, did we do what we needed to get done to protect Lincoln and his generation? Or did we let it slip past us in an irrevocable fashion?

So what is the cost? The cost is the long-term survival of the human race. That is the cost. And the threat is existential.

And this is the greatest country in the world. We should be leading on energy policy, not defining it by the worst polluters on the planet.

So I think this isn't science fiction to get to these levels. I think we already know what we have to do, a mix of nuclear, solar, wind, hydro, and perhaps biofuels. Imagine utilities adopting all this. Electric plug-in cars, and trucks, and ships, and planes, and trains running on it. That we resolve the energy storage crisis with a massive energy efficiency effort.

So I want to ask each of you all in a yes-or-no question: If we gave you the resources with that mix, could we get to the 45 percent drop?

First, Dr. Ekwurzel, could we get there?

Dr. EKWURZEL. If we start now, it is a challenge but we have a chance.

Mr. SOTO. I also want to ask Mr. Williams, could we get there if we had the resources with that mix?

Mr. WILLIAMS. We need to start now.

Mr. SOTO. Reverend Woodberry, do you think it would be possible?

Reverend WOODBERRY. Possibly, but we must start now.

Mr. SOTO. With the Lord's help, right?

Reverend WOODBERRY. Absolutely.

Mr. SOTO. And Congress' help.

And, Mr. Duke, do you think we could do that with that mix?

Mr. DUKE. We could get it done, and could get it done cheaper and faster with a broader mix.

Mr. SOTO. Mr. Powell, would it be possible with that mix?

Mr. POWELL. I would second the broader mix getting it done cheaper and faster.

Mr. SOTO. And then, Mr. Worthington, with the mix I referred to, could we get it done?

Mr. WORTHINGTON. I think you would have to add carbon capture and storage to the technologies that you suggested.

Mr. SOTO. OK. Well, thanks for your opinions on that.

It is my belief the only resource we really need is the will of this committee to meet the challenge of climate change now for Lincoln and his generation. And I believe we have been elected to do just that.

With that, thank you, Lincoln, for being here today. Look at that. See, he has got his political career starting today. And I yield back, Chairman.

Mr. TONKO. The gentleman yields back.

Lincoln and I have met in the past. And, Lincoln, it is great to have you here again. And thank you for being super inspiration.

Now to the very patient Representative Castor from Florida. We offer you 5 minutes to question the panel.

Ms. CASTOR. Thank you, Chairman Tonko and Ranking Member Shimkus. I look forward to tackling these issues with you. Thank you to all of our witnesses.

We are facing the crisis of our generation. The climate crisis threatens all of our districts, all of our communities, as well as America's national security, our economic prosperity, the health of our families, and the world that our children will inhabit. I appreciated my colleague from Florida Mr. Soto's remarks. We feel like we are in the bullseye in Florida.

And my district in the Tampa Bay area is one of the most vulnerable in the country to the impacts of climate change. Hotter and longer summers, deadly storm surge risk because of rising sea levels, more intense hurricanes. It is all impacting the water we drink and even down to the stormwater and wastewater systems that we all rely on every day.

But we are not alone. This is impacting everyone across America. And the costs are very high. Chairman Tonko and I have often talked about the costs of inaction. And right now people are bearing the brunt of higher property insurance costs, flood insurance costs, electric bills. The list goes on and on.

But the good news is there are solutions. We have seen major advances in energy efficiency, renewable energy, innovation, and other strategies to reduce greenhouse gases. The Fourth Climate Assessment Report said that future risks from climate change depend primarily on decisions made today. And it has been heartening to hear some of our Republican colleagues talk about a new understanding of what is at stake.

But, you see, the time is short. The time is short now to avoid the worst impacts and the escalating costs of the climate crisis.

And to my colleague Mr. Duncan, who kind of symbolizes a lot of the discussion we hear on the other side: No, it is absolutely vital that the Congress and this country provide some bold national policies to get there and to tackle the challenges ahead. We have got to tackle the challenges of reducing greenhouse gases, especially in the electric generation sector and transportation sector.

So, to close out, I would like Mr. Duke and Dr. Ekwurzel to talk to us a little bit about that. In the past decade, the average costs of wind and solar electric systems have dropped dramatically and the markets are rapidly growing. With your best can-do spirit, talk to us about the opportunities ahead for this country and communities when it comes to clean energy and the jobs we will create with it.

Dr. EKWURZEL. I will be real short on the resiliency aspect, then I will turn it over to Mr. Duke. Because this is really important. When those are senior citizens that are trapped inside the facility after a hurricane because there is no power because it was disrupted, and the fuel supply lines are disrupted, when the storm passes, the sun comes up and the air still is blowing wind, and you can have a renewable, you know, community solar community wind that can get you back up on your feet, and you can be more independent as you deal with the climate impacts.

Mr. DUKE. Thank you, Representative, for the question.

And I just want to underscore how much progress we have made and how much opportunity we have now to cut emissions faster than ever before. The CEO Jim Robo of the largest utility in America predicts that, within a few years, renewables, wind and solar, with storage will be 2 to 4 cents a kilowatt hour and able to broadly compete with conventional power. That is an indication of what we have got in front of us as we seek to electrify all of our end uses, and building, and vehicles, and beyond.

And I also want to note that there is lots of innovation happening in other sectors. The industrial sector is more complicated. It is one that is hard to get your hands around sometimes, but I want to give an indication of what is going on there.

There is a company in Boston that is creating metals out of electricity in a way that can be cost-competitive even for steel down the line. You have got companies that are using CO2 to strengthen cement in buildings in Atlanta and all across the country. And much more coming in terms of CO2 utilization as part of the overall toolkit.

And, of course, we have long known how to cut energy waste. And increasingly what companies are doing is getting into the system so that they can help with demand response, with flexible loads. For example, there is no reason why you have to charge your electric vehicle right now whenever you first plug it in. It is easy to have that respond to the kinds of rate variations that California is now sending to consumers so that you can charge your electric vehicle when the electricity is most plentiful and cheap.

And this is just a small snapshot of the innovation that is happening right now. Much more to come from small modular reactors to carbon capture and storage, precision agriculture. We can and are in many ways still leading on this, but we need the same kind of 90-plus major policies that China has to make sure that our industries can continue to scale with confidence on all these solutions.

Mr. TONKO. The gentlewoman yields back?

Ms. CASTOR. Yes.

Mr. TONKO. You do.

The Chair recognizes the gentleman from Maryland, Representative Sarbanes, for 5 minutes.

Mr. SARBANES. Thank you very much, Mr. Chairman. I appreciate you all being here today. I am sorry I couldn't be here for a good portion of the hearing, but I did get notes. And I know it has been a very rigorous examination of what we need to do in terms of addressing climate change. And I want to thank the chairman for bringing this hearing and bringing attention to these issues.

Having gotten all the questions that you have received and responded to them over the course of the hearing, I invite you to kind of give a wrap-up perspective on what you think will be the most—pick one, two things—the most effective things that we can do in the nearest term to try to address this crisis of climate change.

And I am also particularly interested, Mr. Duke, in your views on what we can do to incentivize progress on this point other than to the detached issues that have been discussed. If you can start with that and then we can have others give a kind of final perspective.

Mr. DUKE. Thank you, Representative, for the closing questions.

I would like to underscore that the United States has been and really remains the most important player on the world stage for dealing with climate change. It really was the United States and China jointly announcing their targets in 2014 to cut emissions, with China committing to peak their emissions for the first time— and they are delivering on that, by the way—that is what kick started the move to the Paris Agreement, and that is the kind of leadership that we had shown historically and can and will show again.

To be in the position to do that, though, we need to have the right incentives in place that are as far-reaching and market-based as possible. The best way to do that is with a price on carbon that is congressionally bipartisan and that reinvests the revenue that comes out of that carbon price in order to create the right infrastructure, from transmission to electric vehicle charging stations, and to do right by the communities that are on the front lines of this transition, whether it is coal communities or low-income communities suffering from pollution today.

And I can tell you that when we do that, not only will we lead on technology and on the diplomatic stage again, but we will also clean up our public health problems in a dramatic way. When you move to clean energy, you clean up everything> You don't just clean up CO2, you clean up all the public health contaminants as well. And I look forward to seeing bipartisan action on a carbon price that makes all that happen and that allows our business to do their job and compete with China and the rest of the world.

Mr. SARBANES. Any other closing observations, this last?

Dr. EKWURZEL. Don't forget the damages of climate change and global emissions. When you stack that up against these low costs per kilowatt that are already happening, invest in the science, invest in the social science. This is big transformation that I think is going to be a cleaner, healthier world ahead when we act now.

Mr. WILLIAMS. Mr. Sarbanes——

Mr. SARBANES. Yes.

Mr. WILLIAMS [continuing]. It was mentioned on both sides, the moonshot. And I think it is important to note that the moonshot involved Federal intervention, Federal targets, and date-specific goals that was connected with investments and incentives. We need the same thing for climate change.

Mr. SARBANES. Reverend Woodberry.

Reverend WOODBERRY. Community-based solutions that will provide energy efficiency, renewable demand-side management tools that will create jobs, and also a price on carbon, ensuring that that money goes to communities that have a legacy of abuse and pollution.

Mr. POWELL. I will say I heard broad agreement that climate change is a real and urgent problem that we need to address, that we need much higher-ambition policies than we currently have, that we need a full toolkit of solutions to solve the problem, we can't take anything off the table, and that innovation is a really good place to get started.

Mr. WORTHINGTON. I guess I am last. I would just reiterate that both from an energy production side and the efficiency side, we need all of the above. We need every technology that is economically available. Plus, we can't ignore or take any technologies off the table, both on the supply and the utilization side.

Mr. SARBANES. Thank you all. Mr. Chairman, again, thanks for the hearing. I think we agree that we have to move super aggressively in the direction of the side of the portfolio that has to do with green, sustainable energy. The testimony we received today will help us do that.

I yield back.

Mr. TONKO. Thank you very much. And the gentleman yields back.

I believe that completes the list of Members who chose to question the members of the panel. I do thank, very much thank the witnesses for their participation in today's hearing, my first hearing as chair. So I appreciate your cooperation immensely. Thank you for the great inclusion of ideas and thoughts and opportunities that lie before us. We appreciate it greatly.

I remind Members that, pursuant to committee rules, they have 10 business days to submit additional questions for the record to be answered by the witnesses who have appeared. I ask each witness to respond promptly to any such questions that you may receive.

And then, finally, I request unanimous consent to enter the following documents into the record. They include testimony of Jason Hartke, President of the Alliance to Save Energy, Climate Change in the Great Lakes Region: An assessment of Great Lakes Integrated Sciences; a January 8, 2019, letter from the Alliance to Save Energy that was forwarded to Speaker Pelosi, Leader McCarthy, Senate Majority Leader McConnell, and Senate Minority Leader Schumer; a letter from TechNet; a letter from the Advanced Energy Economy; a slide that was provided today by Representative McKinley in his questioning; and, finally, a presentation of slides by the witnesses that accompanied today's involvement.

[The information appears at the conclusion of the hearing.]

Mr. TONKO. So, with all of that, we again thank everyone for their participation and my colleagues for their interest in the issue. And at this time the subcommittee is adjourned.

[Whereupon, at 1:13 p.m., the subcommittee was adjourned.]

[Material submitted for inclusion in the record follows:]

PREPARED STATEMENT OF HON. DEBBIE DINGELL

Thank you Chairman Tonko and Ranking Member Shimkus, for holding this hearing today to discuss the urgent threat from climate change we all face and the way forward.

Sea levels are rising. Average temperatures are warming. Ice is disappearing at alarming rates. Extreme weather is intensifying and becoming more frequent—from stronger hurricanes to colder winters.

The world's top scientific minds have made it clear: the time for debate is overurgent and decisive action is needed now on a significant scale to address climate change. The will of one city, one county, one State, or one country will not be enough to meet the challenge ahead.

In the Great Lakes, we are already seeing increased variability in lake water levels, more harmful algae blooms, and wildlife habitats adversely impacted, which will continue to negatively affect the region's economy and way of life long-term.

It is critical the United States rejoint the rest of the industrialize world as a member of the Paris Climate Accord and take immediate steps to ensure this Nation is transitioning across all sectors to a carbon-zero economy. Repealing, rolling back, or weakening the Clean Power Plan, Clean Air Act, clean car standards, or any other effort to reduce greenhouse gases only exacerbates the climate crisis we need to solve.

With 2018 listed as one of the hottest years on record, the American people have demanded immediate action. Allowing greater climate pollution threatens our public health, our economy, and our national security.

We need bold, new ideas to create a pathway to a clean energy future and create new, good-paying jobs at the same time. We need to make the necessary investments in infrastructure, workforce, and education to mitigate, adapt, and reverse the growing climate threat.

Thank you to all the witnesses for being here today at this important hearing. I am looking forward to working with my colleagues—Republicans and Democrats— on the Energy and Commerce Committee to take serious action and pass meaningful climate legislation this Congress.

We must have the courage to act—the consequences of inaction are real, and all future generation are put at risk each day we do nothing.

WRITTEN TESTIMONY OF JASON HARTKE, PRESIDENT, THE ALLIANCE TO SAVE ENERGY, U.S. HOUSE OF REPRESENTATIVES, ENERGY AND COMMERCE COMMITTEE, SUBCOMMITTEE ON ENVIRONMENT AND CLIMATE CHANGE, "TIME FOR ACTION: ADDRESSING THE ENVIRONMENTAL AND ECONOMIC EFFECTS OF CLIMATE CHANGE," FEBRUARY 6, 2019

Thank you for the opportunity to submit a written statement regarding the Environment & Climate Change Subcommittee's Feb. 6 hearing titled, "Time for Action: Addressing the

Environmental and Economic Effects of Climate Change." We are pleased that the Committee is taking a close look at the consequences of climate change and the actions that must be taken to address it. We look forward to working with you to develop and advance bipartisan climate policy in the 116th Congress.

I write today to emphasize the powerful role that energy efficiency can play as a solution, and the bipartisan opportunity it represents. Energy efficiency is widely recognized as the single most effective strategy we have for reducing carbon emissions, with the International Energy Agency reporting that efficiency alone must account for more than 40 percent of the emissions reductions needed to meet global targets. At the same time, increased efficiency is an enormous economic opportunity, with tremendous untapped potential to create jobs and economic activity – particularly in the construction and manufacturing sectors – while strengthening U.S. productivity and competitiveness and saving Americans billions of dollars in energy costs.

Energy efficiency already is by far the largest sector in the clean energy industry, supporting more than 2.25 million jobs (out of a total of 3.1 million clean energy jobs). It is also among the fastest-growing, with employers anticipating nearly 10 percent annual growth. Seven in 10 of those jobs are in construction and manufacturing sectors. Encouraging efficiency will stimulate economic activity across the country and drive innovation and technology development that will help maintain U.S. leadership in an increasingly competitive global industry.

As you search for bipartisan solutions in the 116th Congress, we urge you to start with bipartisan policy encouraging energy efficiency, whether through infrastructure and transportation policy, tax incentives, public-private partnerships, R&D investments, or other policy areas. *Attached, please find a letter dated Jan. 8 signed by more than 40 companies and organizations also calling on Congress to prioritize efficiency.*

Opportunities for advancing energy efficiency exist across a broad range of policy areas, including:

- Infrastructure: Infrastructure is more than roads and bridges – it's our utility grid, water and wastewater facilities, transit hubs, public buildings, ports, and other structures. These facilities use enormous amounts of energy, and a nationwide infrastructure initiative presents an opportunity to "get it right" and save consumers and taxpayers decades of wasted energy costs. In some cases, infrastructure projects can pay for themselves through public-private partnerships and innovative financing of energy savings investments. Incorporating energy efficiency can also provide a host of additional benefits, such as improving power grid reliability and resilience by stabilizing demand and reducing emissions – all while creating good-paying jobs.
- Transportation: The U.S. transportation sector – which accounts for about one-third of U.S. energy consumption and carbon emissions – is on the verge of a major transformation that has enormous implications for energy use. New technologies and business models such as ride-sharing, electrification, autonomous

vehicles, and data-driven public transportation are creating an opportunity to reinvent mobility for a smarter, more integrated system that uses energy more efficiently. This requires new policy and coordination. In 2017, the Alliance convened the 50x50 Commission on U.S. Transportation Sector Efficiency, including automakers, utilities, public interest groups, product manufacturers, and technology providers. With a goal of reducing transportation energy use by 50 percent by 2050, the Commission in September outlined a series of recommendations including expanding tax incentives for electric and other alternative-fuel vehicles, investing in charging infrastructure, maintaining strong fuel efficiency standards, investing in research, development and deployment to strengthen U.S. leadership, and improving coordination among different jurisdictions.

- Built Environment: Existing homes and buildings – and new ones under construction – will be in use for decades to come, with enormous implications for U.S. energy consumption. The built environment currently accounts for about 40 percent of our energy use, and as with the transportation sector, innovation and technology are creating new opportunities for savings in residential, commercial, and industrial applications that can play a significant role in decarbonizing the economy. In addition to encouraging traditional efficiency solutions such as improved building envelopes and equipment, there are tremendous policy opportunities to pave the way for highly efficient homes and buildings through systems-oriented practices and technologies such as integrated design, active-energy management, internet of things, grid integration, and artificial intelligence.
- Tax Policy: While the federal government encourages nearly every mainstream form of energy generation with tax incentives – and has done so for decades – there are currently no direct incentives for energy efficiency in the U.S. tax code. This is a glaring and costly omission. Opportunities for encouraging high-efficiency homes and buildings could lock in decades of energy and cost

savings while stimulating construction activity. Likewise, long-term, meaningful incentives for high-efficiency air conditioning, water heaters, lighting systems, and other equipment are proven to stimulate markets, save consumers money, and sharply reduce emissions.

- Federal Program Funding: Federal investments in energy efficiency drive gains throughout the economy and stimulate billions of dollars in economic activity. Third-party, peer-reviewed studies show that total taxpayer investment of $12 billion to date in R&D at the Department of Energy's Office of Energy Efficiency and Renewable Energy has yielded more than $388 billion in net U.S. economic benefits. Public private partnerships such as the ENERGY STAR and Better Plants/Better Buildings Initiative have delivered equally impressive results. For example, in addition to helping Americans save more than $30 billion in 2016, ENERGY STAR saved 400 billion kWh of electricity, delivering carbon emission reductions of 320 million metric tons – or the equivalent of the entire energy use of nearly 35 million homes for one year.

The U.S. has come a long way in using energy more productively in recent decades, yet more than half of the energy we generate is still not put to use. The opportunities ahead are even greater than our past gains, and as a diverse coalition of businesses and organizations, we stand ready to work with you to continue U.S. leadership in this important field and advance efficiency across the economy.

About the Alliance to Save Energy

Founded in 1977, the Alliance to Save Energy is a nonprofit, bipartisan alliance of business, government, environmental and consumer leaders working to expand the economy while using less energy. Our mission is to promote energy productivity worldwide – including through energy efficiency – to achieve a stronger economy, a cleaner environment and greater energy security, affordability and reliability.

Sincerely,
Jason Hartke
President
The Alliance to Save Energy

Climate Change in the Great Lakes Region

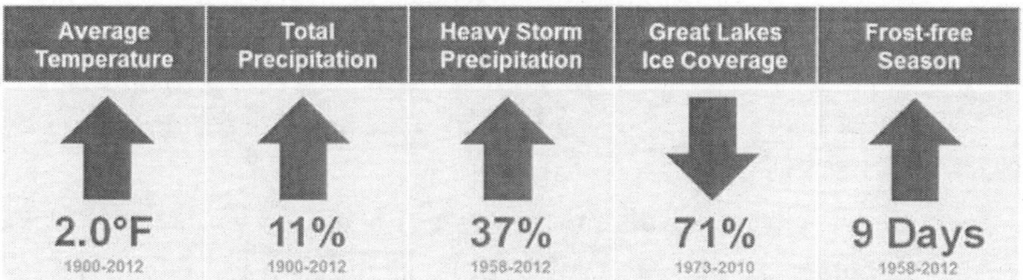

Temperature

- Since 1900, annual average temperatures have increased by 2.0°F (1.1°C) in the U.S. Great Lakes region.
- By 2050, average air temperatures are projected to increase by 1.8 to 5.4°F (1 to 3°C).
- By 2100, average air temperatures are projected to increase by 3.6 to 11.2°F (2 to 6.2°C).

Precipitation

- Since 1900, total annual precipitation has increased by 10.8% in the U.S. Great Lakes Region, and is expected to continue to increase, though projections of future precipitation vary.

- Precipitation will increase during wet seasons but may remain nearly stable or decrease during the summer.
- Reduced lake ice coverage will result in more exposed water and more opportunity for lake-effect precipitation.

Snow, Ice Cover and Lake Temperature

- Lake temperatures have been increasing faster than surrounding air temperatures.
- From 1973 to 2010, annual average ice coverage on the Great Lakes declined by 71%.
- From 1975 to 2004, the annual number of days with land snow cover decreased by 15 and the average snow depth decreased by 2 inches (5.1 cm).
- Snow and ice levels on the Great Lakes and on land will likely continue to decrease, with little significant ice cover on Lake Superior by mid-century in a typical year.

Extreme Weather

- The frequency and intensity of severe storms has increased. This trend will likely continue as the effects of climate change become more pronounced.
- The amount of precipitation falling in the heaviest 1% of storms increased by 37% in the Midwest and 71% in the Northeast from 1958 through 2012.
- More severe storms may have a negative economic impact due to resulting damages and increased costs of preparation, clean up, and business disruption.

Water Quality and Stormwater Management

- Increased risk of droughts, severe storms, and flooding events may increase the risk of erosion, sewage overflow, lead to more interference with transportation, and more flood damage.
- Future changes in land use could have a far greater impact on water quality than climate change. The coupling of climate change and land use change could therefore result in even stronger effects in some areas.

Lake Levels

- Long-term water levels in the Great Lakes have fallen since reaching record highs in the 1980s.
- While most models project continued, long-term declines in lake levels, shorter-term variations will remain large, and periods of high lake levels are probable.
- Other factors, such as lake regulations, also affect lake levels, though no major management changes have occurred since 2000.

Algal Blooms

- Warmer water surface temperatures increase stratification of the lakes and decrease vertical mixing.
- Stronger storms and the use of impervious surfaces increase runoff and nutrient loading to the Great Lakes.
- Combined sewer overflows and agricultural fertilizers are major contributors to high nutrient loads.
- Stronger storms, warmer temperatures, and nutrient loading are conspiring to produce more hypoxic dead zones and toxic algal blooms.

Fish and Wildlife

- The rate of warming may outpace the rate at which ecosystems are able to migrate and adapt.
- Wildlife populations better adapted to cold temperatures will continue to decline as competing species migrate into the region from the south with rising temperatures.
- Lake stratification and more hypoxic conditions will further stress biomass productivity in lakes and wetlands.
- Increased evaporation rates may decrease wetland area in the region.

Energy and Industry

- Reduced summer water availability may interfere with some industrial operations.
- Warmer temperatures and more frequent heat waves will likely increase electricity demands, particularly in urban areas and during the summer months.

Forests

- As temperatures rise, the distribution and composition of tree species will change and shift northward.
- With warmer temperatures and increasing CO_2, forest productivity will likely increase until other impacts of climate change, such as increased drought, fire, and invasive species present additional stressors to forests.

Water Availability

- Despite increasing precipitation, land surfaces in the Great Lakes region are expected to become drier overall due to increasing temperatures and evaporation rates.
- More frequent summer droughts could affect soil moisture, surface waters, and groundwater supply.
- The seasonal distribution of water availability will likely change. Warmer temperatures may lead to more winter rain and earlier peak streamflows.

Agriculture

- The frost-free season lengthened by 9 days in the Midwestern U.S. and 10 days in the Northeast from 1958-2012, and may be up to 1-2 months longer by 2100.
- Through mid-century, a longer growing season and higher CO_2 concentrations will likely have a positive effect on many crop yields.
- By 2100, the negative effects of increasing storm activity, flooding, extreme heat, summer drought risk, and pests may outweigh the benefits of other climate changes.

Transportation

- More extreme heat may increase the risk of heat damage to pavement and rails.
- More extreme precipitation may compromise transportation routes and damage infrastructure.
- Shipping lanes will likely be open earlier and longer due to reduced ice cover on the Great Lakes.

- Lower lake levels lead to decreased depth of navigation channels and a reduction in the maximum loads carried by vessels. For each inch of lost draft, the average 1,000-foot freighter loses $30,000 per transit.

Public Health

- Increased risk of heat waves and increased humidity may increase the number of heat-related deaths and illnesses.
- More storm activity and flooding will increase the risk of watershed contamination while warmer surface waters amplify the risk of toxic algal blooms and fish contamination.
- Diseases such as West Nile virus and Lyme disease may become more widespread since carrier insects will be more likely to survive milder winters.

Tourism and Recreation

- Winter recreation and tourism are likely to suffer due to reduced snow cover and shorter winters.
- Increased lake contamination and decreasing lake levels may lead to less desirable shorelines, but increasing summer temperatures and a longer summer season, may increase demand for beaches.
- Overall, summer tourism may grow before temperatures rise become unfavorable for many recreational activities.
- Many coldwater species of fish important to recreation are likely to decline while populations of warmwater species grow.

January 8, 2019

The Honorable Nancy Pelosi
Speaker
U.S. House of Representatives

The Honorable Mitch McConnell
Majority Leader
U.S. Senate

The Honorable Kevin McCarthy
Minority Leader
U.S. House of Representatives

The Honorable Charles E. Schumer
Democratic Leader
U.S. Senate

Dear Speaker Pelosi, Leader McConnell, Leader Schumer, and Leader McCarthy,

Climate change is becoming a defining issue of our time as its effects become more widely known and studies show it is accelerating faster than anticipated. Recent reports, including the Fourth National Climate Assessment released by the Trump administration, highlight the longterm economic costs and environmental consequences facing the United States. Momentum is growing on both sides of the aisle for addressing this challenge in a way that strengthens U.S. economic productivity and competitiveness.

As members of the Alliance to Save Energy – businesses and organizations representing thousands of American workers in energy efficiency – we write to remind policymakers that energy efficiency is one of the most effective strategies we have for addressing this growing threat, representing an extraordinary bipartisan opportunity to reduce carbon emissions while simultaneously boosting economic growth and job creation, strengthening U.S. leadership in innovation, improving our energy security, and advancing global competitiveness. As you search for bipartisan solutions in the 116th Congress, we urge you to seize this opportunity by prioritizing energy efficiency, whether through infrastructure and transportation policy, tax incentives, public-private partnerships, R&D investments, or other initiatives.

The benefits of energy efficiency are clear and compelling:

- Studies have repeatedly found that energy efficiency is the most effective strategy for reducing carbon emissions. The International Energy Agency reports that efficiency alone can account for more than 40 percent of the emissions reductions needed to meet global targets. The agency and many other experts recommend that climate strategy should begin with energy efficiency policy that reduces energy consumption in buildings, transportation, manufacturing, and equipment such as heating and air conditioning and lighting systems.
- Energy efficiency is by far the largest sector in the clean energy industry, supporting more than 2.25 million jobs (out of a total of 3.1 million clean energy jobs). It is also among the fastest-growing, with employers anticipating nearly 10 percent growth in 2018. Energy efficiency is also an important tool in creating a more productive and competitive U.S. economy. Advanced manufacturing practices, for example, can play a key role in rebuilding the American manufacturing sector, while efficiency technologies represent a large and growing global market.
- Efficiency gains resulting from federal policies and programs save consumers billions of dollars in energy costs annually – money that consumers can plough back into the economy. The average household, for example, saves almost $500 every year from federal efficiency standards for common appliances such as dishwashers, dryers, and water heaters, while ENERGY STAR – which has a $42 million budget – helped Americans save more than $30 billion in energy costs in 2016 alone.

Opportunities for advancing energy efficiency exist across a broad range of policy areas, including:

- Infrastructure: Infrastructure is more than roads and bridges – it's our utility grid, water and wastewater facilities, transit hubs, public buildings, ports, and other structures. These facilities use enormous amounts of energy, and a nationwide infrastructure initiative presents an opportunity to "get it right" and save consumers and taxpayers decades of wasted energy costs. In some cases, infrastructure projects can pay for themselves through public-private partnerships and innovative financing of energy savings investments. Incorporating energy efficiency can also provide a host of additional benefits, such as improving power grid reliability and resilience by stabilizing demand and reducing emissions – all while creating good-paying jobs.
- Transportation: The U.S. transportation sector – which accounts for about one-third of U.S. energy consumption and carbon emissions – is on the verge of a major transformation that has enormous implications for energy use. New technologies and business models such as ride-sharing, electrification, autonomous vehicles, and data-driven public transportation are creating an opportunity to reinvent mobility for a smarter, more integrated system that uses energy more efficiently. This requires new policy and coordination. In 2017, the Alliance convened the 50x50 Commission on U.S. Transportation Sector Efficiency, including automakers, utilities, public interest groups, product manufacturers, and technology providers. With a goal of reducing transportation energy use by 50 percent by 2050, the Commission in September outlined a series of recommendations including expanding tax incentives for electric and other alternative-fuel vehicles, investing in charging infrastructure, maintaining strong fuel efficiency standards, investing in research, development and deployment to strengthen U.S. leadership, and improving coordination among different jurisdictions.

- Built Environment: Existing homes and buildings – and new ones under construction – will be in use for decades to come, with enormous implications for U.S. energy consumption. The built environment currently accounts for about 40 percent of our energy use, and as with the transportation sector, innovation and technology are creating new opportunities for savings in residential, commercial, and industrial applications that can play a significant role in decarbonizing the economy. In addition to encouraging traditional efficiency solutions such as improved building envelopes and equipment, there are tremendous policy opportunities to pave the way for highly efficient homes and buildings through systems-oriented practices and technologies such as integrated design, active-energy management, internet of things, grid integration, and artificial intelligence.
- Tax Policy: While the federal government encourages nearly every mainstream form of energy generation with tax incentives – and has done so for decades – there are currently no direct incentives for energy efficiency in the U.S. tax code. This is a glaring and costly omission. Opportunities for encouraging high-efficiency homes and buildings could lock in decades of energy and cost savings while stimulating construction activity. Likewise, long-term, meaningful incentives for high-efficiency air conditioning, water heaters, lighting systems, and other equipment are proven to stimulate markets, save consumers money, and sharply reduce emissions.
- Federal Program Funding: Federal investments in energy efficiency drive gains throughout the economy and stimulate billions of dollars in economic activity. Third-party, peer-reviewed studies show that total taxpayer investment of $12 billion to date in R&D at the Department of Energy's Office of Energy Efficiency and Renewable Energy has yielded more than $388 billion in net U.S. economic benefits. Public private partnerships such as the ENERGY STAR and Better Plants/Better Buildings Initiative have delivered equally impressive results. For example, in addition to

helping Americans save more than $30 billion in 2016, ENERGY STAR saved 400 billion kWh of electricity, delivering carbon emission reductions of 320 million metric tons – or the equivalent of the entire energy use of nearly 35 million homes for one year.

The U.S. has come a long way in using energy more productively in recent decades, and it should continue to take a global leadership position. The opportunities ahead are even greater than our past gains, and as a diverse coalition of businesses and organizations, we stand ready to work with you to advance energy efficiency throughout the U.S. economy.

Sincerely,

A.O. Smith
Alliance for Water Efficiency Alliance to Save Energy
American Association of Blacks in Energy
American Council for an Energy-Efficient Economy
American Institute of Architects
American Public Transportation Association
Andersen Corporation
Association of Energy Engineers
Austin Energy
California Energy Commission
Copper Development Association
Daikin US
Danfoss
DFW International Airport
The Dow Chemical Company
DuPont
Hannon Armstrong
Illuminating Engineering Society
Ingersoll Rand
Intel
International Window Film Association

Johnson Controls
Knauf Insulation
Legrand
Lime Energy
Midwest Energy Efficiency Alliance
National Grid
Natural Resources Defense Council
New York Power Authority
North American Insulation Manufacturers Association
Pacific Gas and Electric Company
Polyisocyanurate Insulation Manufacturers Association
Sacramento Municipal Utility District
Schneider Electric
Seattle City Light
Sense
Signify
Southern California Edison
U.S. Green Building Council
VEIC

CC: Members of the 116th Congress

February 6, 2019

The Honorable Paul Tonko
Chairman
House Energy and Commerce Committee Environment and Climate Change Subcommittee

The Honorable John Shimkus
Republican Leader
House Energy and Commerce Committee Environment and Climate Change Subcommittee

Dear Chairman Tonko and Republican Leader Shimkus:

We welcome this opportunity to submit comments for the official record of the February 6, 2019 hearing by the House Energy and Commerce Committee Environment and Climate Change Subcommittee titled, "Time for Action: Addressing the Environmental and Economic Effects of Climate Change."

TechNet is the national, bipartisan network of innovation economy CEOs and senior executives. Our diverse membership includes dynamic American businesses ranging from startups to the most iconic companies on the planet and represents over three million employees and countless customers in the fields of information technology, e-commerce, the sharing and gig economies, advanced energy, cybersecurity, venture capital, and finance.

While we recognize there are disagreements among policymakers about the best solutions to tackle climate change, we believe growing bipartisan support for action can lead us to find some common ground and achieve meaningful progress during this Congress.

Throughout our 22-year history, we have championed policies at the federal and state levels that support advanced energy innovation and efforts to address climate change. Our member companies are also committed to combatting climate change and are investing significantly in innovations and initiatives that not only lower their own carbon footprint but also help other companies across different sectors deploy advanced energy solutions to achieve the same goal.

Furthermore, the following statements are enshrined in our federal policy principles: "TechNet supports advanced energy policies that foster and promote a climate for innovation. TechNet supports market-based policies that enable companies to create, thrive, and participate in competitive energy markets in the United States and around the globe, and balanced action on climate change that accelerates the deployment of low and zero-carbon energy technologies."

Our words have been backed by actions and policy outcomes that have accelerated the shift to advanced clean energy solutions that are making a meaningful difference in addressing global climate change. For example, we have successfully advocated for policies increasing the adoption of electric vehicles (EV) by encouraging innovation and investment in a nationwide EV charging infrastructure; promoting the development of solar, wind, fuel cells, storage, and other forms of clean energy generation; and ensuring that consumers can purchase clean energy to lower emissions, protect the environment, and reduce the load on the electrical grid.

In 2017, TechNet successfully urged the Federal Energy Regulatory Commission (FERC) to reject a proposed rule that would have severely disrupted energy markets in favor of coal. According to *The Economist*, "If the FERC adopted [the] rule, it would have amounted to one of the biggest government interventions in energy markets for decades, and risks frightening investors by putting the thumb on the scale for coal and introducing policy uncertainty." Instead, the rejection of this rule in January 2018 was an important step to promote stability and affordable pricing in power markets.

In recent years, TechNet has also been a proud member of a broad coalition working to secure annual funding for the Department of Energy's Advanced Research Projects Agency – Energy (ARPA-E) program, which plays a unique and critical role in maintaining America's global leadership in energy technologies. TechNet also supported restoring the Section 48 Investment Tax Credit as part of the 2018 bipartisan budget bill. This important provision allows project owners or investors to qualify for federal business energy investment tax credits for installing designated renewable energy generation equipment placed in service between 2006 through 2024. Since it was first enacted, this policy has helped drive the growth of the clean energy industry and reduced emissions in the electricity sector to 14 percent below 2005 levels.

Although we have achieved several policy victories, we also recognize that federal policymakers are not prioritizing these issues and have even taken steps in the wrong direction. For example, as an industry, we were disappointed the Trump Administration abandoned the Paris Climate Accords in 2017. Since then, evidence of climate change's threats has only increased.

For example, in November 2018, the Trump Administration's own National Climate Assessment reaffirmed what we have known for some time: that inaction on climate change will harm our economy. In addition to having environmental and economic consequences, climate change poses national security challenges as evidenced by the Department of Defense's January 2019 report showing that at least two-thirds of the U.S. military's installations are threatened by flooding, drought, and wildfires driven by climate change over the next two decades alone. This is particularly concerning to us given the tech industry's longstanding partnership with the military in advancing U.S. national security objectives.

In sum, addressing climate change is a priority for TechNet and the industry we represent. Our membership includes not only advanced energy technology innovators, but also some of the largest energy consumers in the world. These companies have invested in innovative technologies to meet their environmental and sustainability goals for their organizations, employees, and consumers, and their need for reliable, uninterrupted energy to power their data and transaction centers and other facilities. Promoting these innovations at both the federal and state levels is not only in the interest of protecting our environment, it is in our economic and national security interests as well.

Thank you for considering our perspective as you hold this important hearing. We welcome the opportunity to serve as a resource to the subcommittee, the full committee, and the U.S. House of Representatives as you and your colleagues continue examining this important issue.

Sincerely,

Linda Moore
TechNet President and CEO

CC: The Honorable Frank Pallone, Jr., Chairman, House Energy and Commerce Committee

The Honorable Greg Walden, Republican Leader, House Energy and Commerce Committee

February 5, 2019

The Honorable Frank Pallone
Chairman
Energy and Commerce Committee
U.S. House of Representatives

The Honorable Greg Walden
Ranking Member
Energy and Commerce Committee
U.S. House of Representatives

The Honorable Paul Tonko
Chairman
Environment & Climate Change Subcommittee
U.S. House of Representatives

The Honorable John Shimkus
Ranking Member
Environment & Climate Change Subcommittee
U.S. House of Representative

Re: Hearing assessing the environmental and economic impacts of climate change

Dear Chairman Pallone, Ranking Member Walden, Chairman Tonko, and Ranking Member Shimkus,

On behalf of the advanced energy industry, we write to offer support as you consider legislative action to address the risks posed by climate change. The recent report by the National Climate Assessment demonstrated the economic risks and potential adverse impacts of climate change to our country. As your committee considers actions that government can take in response, we encourage you to look at advanced energy technologies and services. Relying on these technologies will reduce the risks associated with climate change, while at the same time increase the affordable energy options available to businesses and consumers, improve the reliability and resilience of the grid, and create jobs across the country.

Today, the $200 billion advanced energy industry makes up a strong segment of the American economy, supporting more than 3 million jobs across the country. Increasingly, advanced energy technologies and services save consumers billions of dollars. Costs have fallen so sharply that in some parts of the country investing in new wind and solar energy projects are more cost--effective than continuing to operate existing fossil fuel power plants. For example, a recent utility filing by Northern Indiana Public Service found customers could save $4 billion by replacing its entire existing coal fleet by 2028 with a portfolio of solar, wind, storage, and demand management resources. Investments in advanced energy provide opportunities for job growth and lower electricity costs to consumers while also lowering carbon emissions across the economy.

To maximize the benefits to the economy, consumers, and environment, we encourage you to put to work advanced energy technologies in any federal response to climate change. Advanced energy encompasses the best available energy technologies for supply and demand, including wind, solar, energy storage, nuclear, demand response, energy efficiency, electric vehicles, hydropower, combined heat and power, and fuel cells among others.

We support calls for action to increase the use of certain advanced energy technologies rapidly by 2030, but evidence demonstrates the most cost--effective way to achieve climate goals is by using a broad suite of advanced energy technologies and services. Investment in advanced energy technologies can achieve multiple outcomes, including job creation, protecting communities, increasing resilience, and prioritizing fairness and economic opportunity for those most affected by climate impacts.

We also understand your interest and support your calls for legislative action to improve the nation's infrastructure as a key part of addressing the risks posed by rising carbon emissions. A comprehensive infrastructure plan must include the use of broad advanced energy technologies and services including transmission expansion, microgrids, digital solutions, and transportation electrification strategies.

We support legislative action to deploy advanced energy technologies and services to address the impacts of carbon emissions on the environment, public health, and the economy. We look forward to working with all members of the 116th Congress on this opportunity.

Sincerely, Nat Kreamer
Chief Executive Officer
Advanced Energy Economy

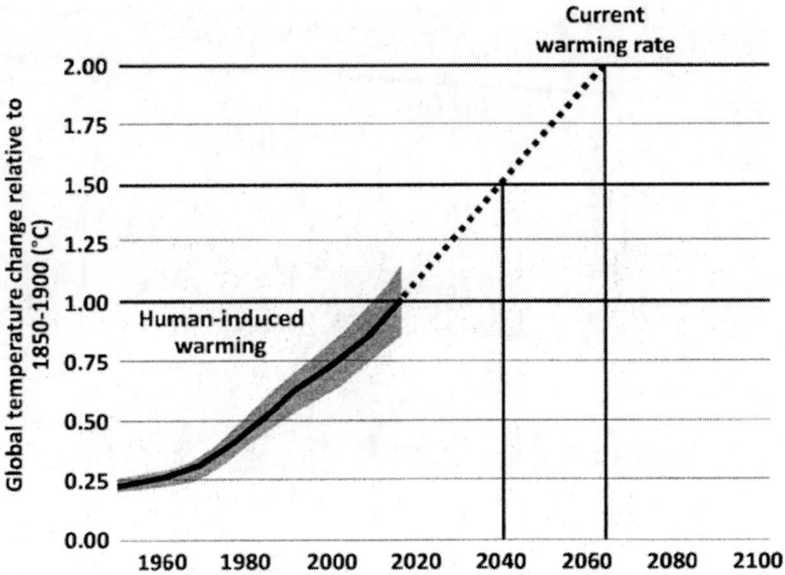

IPCC 2018 SR15 Fig FAQ 1.2

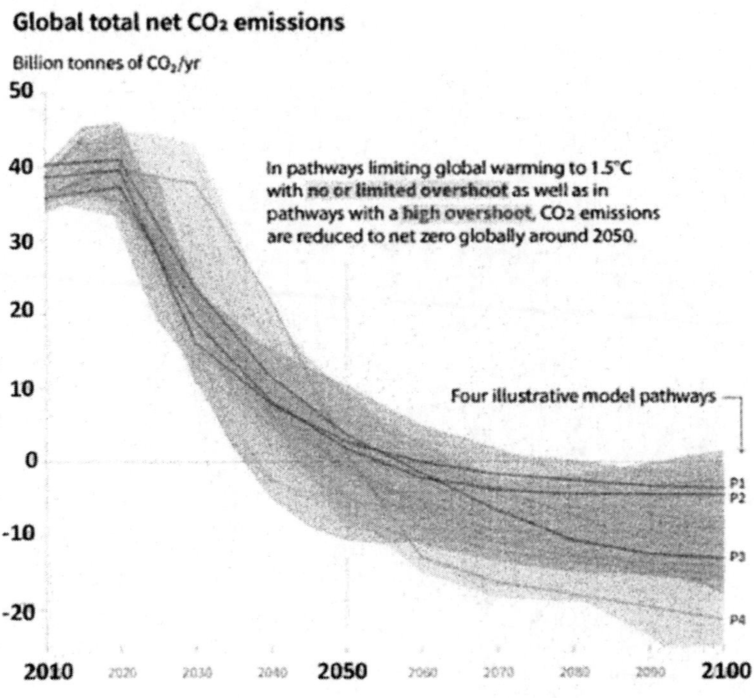

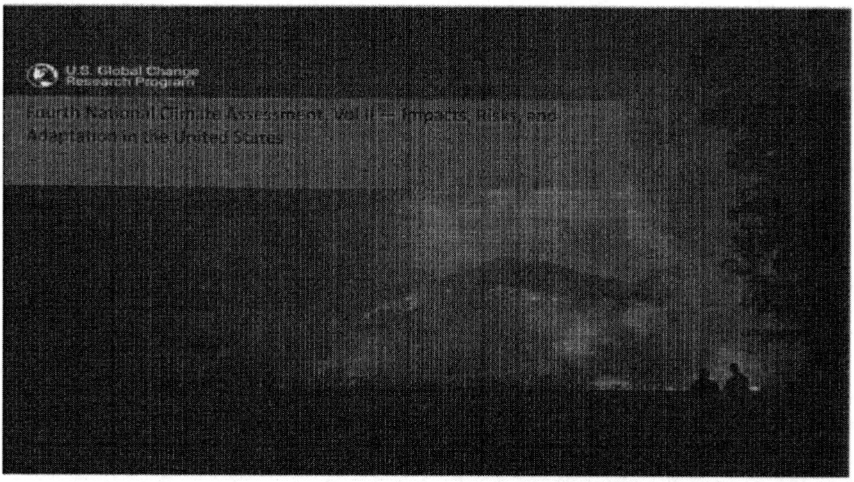
U.S. Global Change
Research Program
Fourth National Climate Assessment, Vol II — Impacts, Risks, and
Adaptation in the United States

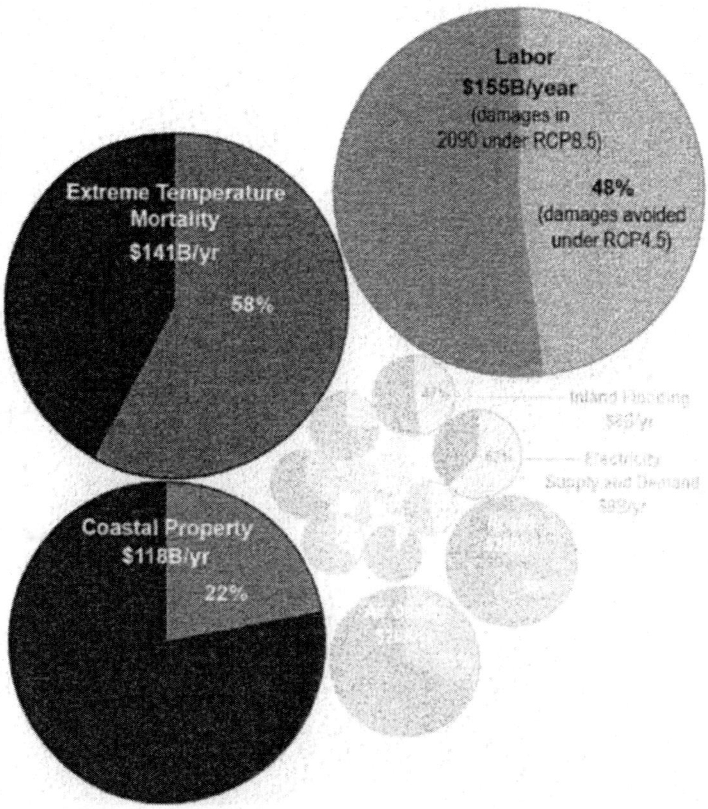
Labor
$155B/year
(damages in
2090 under RCP8.5)
48%
(damages avoided
under RCP4.5)
Extreme Temperature
Mortality
$141B/yr
58%
Coastal Property
$118B/yr
22%
Inland Flooding
Electricity
Supply and Demand

Mitigation-related activities across the U.S.

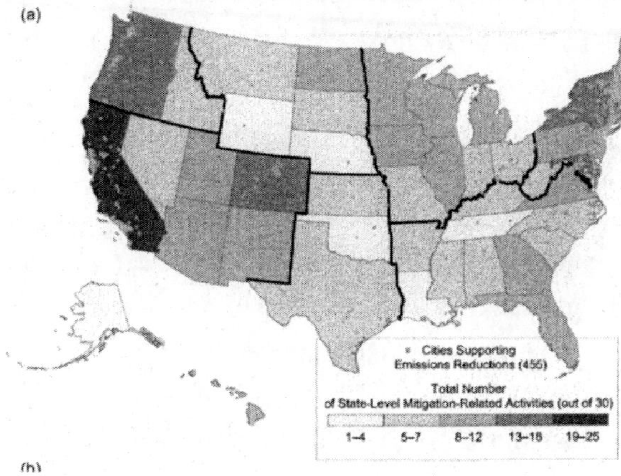

EPA and ERT, Inc.

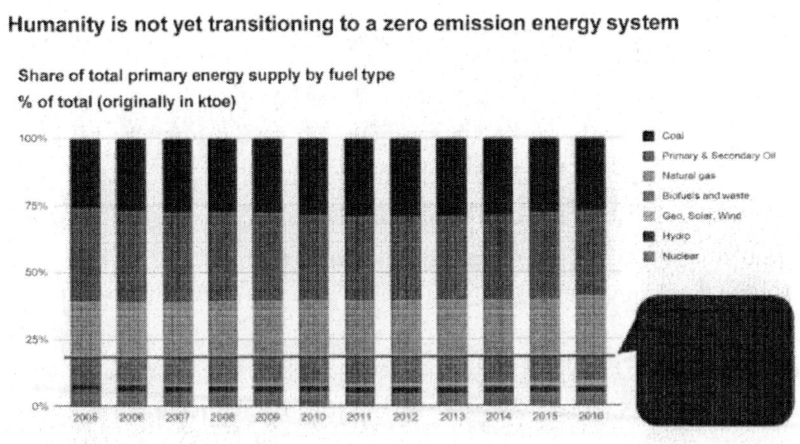

Source: International Energy Agency World Energy Balances 2018

From: Richard J. Powell, Executive Director, ClearPath

Due: March 25, 2019

The Honorable Markwayne Mullin (R-OK)

The Democrats Green New Deal plans on getting rid of fossil fuels in 10 years. With the United States producing only roughly 18% of electricity from renewable sources right now, how would we replace the 62% that comes from natural gas and coal?

a. Can you talk more about how carbon capture could help us reduce CO2 emissions while still using the fossil fuels our country relies on?

No credible analysis shows the United States can completely transition from fossil fuels in a ten year period. The Green New Deal would require renewable energy to be adopted across all aspects of our economy faster than the most popular consumer products in recent history, including the adoption of cellphones and flat-screen TVs. The proposal is made even more daunting because fossil fuels are what (quite literally) moves and powers the American economy. Natural gas and coal make up 62% of the United States electricity mix.

Economy-wide, including sectors like transportation and heavy industry, fossil fuels make up 80% of total U.S. energy use.[88] Putting costs and economics aside, complete replacement of these fuels in a decade is highly improbable.

Carbon capture can help reduce carbon dioxide emissions from fossil fuel consumption in a very significant way, in line with both economic and environmental goals. According to groups as diverse from the largest energy companies (Shell, Southern Company, etc.), leading universities (MIT, Stanford, etc.), and international energy & environmental authorities (International Energy Agency, Intergovernmental Panel on Climate Change) carbon capture can be a very significant lever for reducing global carbon emissions. For several decades, carbon capture has been applied to enhanced oil recovery where we can now can capture over 25 million tons of industrial carbon dioxide each year.[89] The most U.S. recent facility, Petra Nova, captures carbon dioxide from a coal power plant. It can capture 90% of all emissions that run through the system, or roughly the equivalent of carbon dioxide produced by 350,000 cars.[90]

---

[88] https://www.eia.gov/energyexplained/?page=us_energy_home.

[89] https://co2re.co/FacilityData.

[90] https://www.nrg.com/case-studies/petra-nova.html.

The Department of Energy is a global leader in carbon capture research and development. They've help drive dramatic cost reductions and have been responsible for several commercial projects in the United States, from Texas to Illinois. In tandem with the recently enacted changes to the Section 45Q carbon capture tax credit, the Department of Energy expects more commercial carbon capture projects.

Unlocking carbon capture is important because it can address carbon emissions from fossil-fired power plants, contributing to the stability and affordability of the power grid. Additional research should also be conducted to improve the performance of carbon capture systems at more industrial sources, including steel and cement facilities. To realize this promising reality, Congress should encourage R&D across the development spectrum and address regulations that restrict the buildout of enabling infrastructure.

To: The Honorable Frank Pallone, Jr. (D-NJ)
Congress of the United States
House of Representatives
Committee on Energy and Commerce

From: The Reverend Leo Woodberry, Executive Director
Justice First
Florence, SC

RE: Questions for The Reverend Leo Woodberry Testimony
At the "Time for Action: Addressing the Environmental and Economic Effects of Climate Change"
Date: May 1, 2019

Question 1: What role should community-based energy solutions (such as community solar) play in addressing climate change at the local level?

Answer: Community-based climate change solutions will vary from location to location, depending on conditions and situations. Of course, to mitigate carbon and greenhouse gas emissions, we need to move as rapidly as possible towards generating clean renewable energy, such as solar. Mitigation may also include stopping large scale logging along rivers and waterways, since our forests and trees are our natural defenses against weather-related flooding. In rural areas, resilience may include cleaning out drainage ditches and urban areas, improving and expanding infrastructure for storm water drainage and sewers. In both rural and urban area, we may want to look at bioswells and retention ponds, as both resilience and adaptation solutions. Another increasing concern is to elevate homes throughout the country that are located in expanding and increasing floodplains. Finally, federal subsidies and tax credits for energy efficiency and demand side management devices (such as: programmable thermostats and timers for water heaters) will lower energy costs for utilities, electric co-ops, and customers.

Question 1a: What are some of the benefits that community solar offers the customers it serves?

Answer: The challenges facing many residential customers when it comes to solar generated energy, are as follows: (1) homes that are facing in the wrong direction. (2) Homes that are overshadowed by trees and vegetation. (3) Homes that have old roofs or trailer homes with roof structures that are not strong enough. (4) Residents who live in apartment complexes, whose landlords cannot afford to install solar panels.

The benefits of community shared solar is that, all of the aforementioned customers residing in homes described above, can benefit from solar energy generated from community shared solar farms. They do not have to purchase rooftop systems that may run as high as $30-50,000 and are out of the financial reach of many Americans. Another benefit is that community shared solar provides electricity for low-income customers whose income tax situation will not allow them to benefit from federal and state tax credits.

Question 1b: Please share some examples of where community solar has been most successful.

Answer: A good example of a community shared solar project is provided by Duke Energy in Dillon County, SC. The utility in partnership with the Whitney M. Slater Foundation, Kingdom Living Temple, and

New Alpha Community Development Corporation, constructed a solar farm in the middle of a soybean farm that serves 1,200 households. The solar farm is located in a predominately low-income, rural community. Low-income individuals had their one time $250 connection fee waived for 300 residential customers. The utility also provided 1,500 free energy efficiency upgrades. This will minimize the replication of models of energy inefficiency and injustice. Such a model would have customers heating and cooling the outdoors, spending a disproportionate amount of their income on energy, and costing the utility additional expenses to disconnect and reconnect customers. This inclusive process of involving the community with the planning and implementation of community shared solar has been praised and the South Carolina Solar Council will be awarding the community-based organizations with an innovative solar award this week.

Question 1c: What obstacles have you encountered in developing community solar and how can the federal government help support community-based solutions?

Answer: We have encountered several obstacles in the two years preceding the community shared solar farm, constructed in Dillon, SC. The obstacles included: a cap and sunset provision on net metering and solar generation. The need for subsidies or grants for utilities and co-ops, who are reluctant to construct small scale solar projects because of the length of time it takes for them to get a return on their investment. There is a need for grants that can be awarded to nonprofit community-based organizations, who can meet the resilience, mitigation, and adaptation needs of communities impacted by climate change and to construct small scale solar projects for subdivisions, mobile home parks, and church and school parking lots. These grants and other forms of federal assistance can also provide jobs in rural and low-income urban communities that are desperately in need of new engines of economic development.

Barry K. Worthington, Executive Director, United States Energy Association, Written Responses to Testimony Questions from the Honorable Markwayne Mullin (R-OK), U.S. House Energy and Commerce, Subcommittee on Environment and Climate Change, Hearing, "Time for Action: Addressing the Environmental and Economic Effects of Climate Change," February 6, 2019

Thank you, Congressman Mullin, for these questions and the opportunity to add to the record on this important issue.

Natural gas is at the heart of the energy revolution in the United States. Shale gas and shale oil development have allowed us to go from an era of scarcity to an era of abundance.

1. Does natural gas help the United States reduce its reliance on foreign energy sources?
   Yes, natural gas does help the United States reduce its reliance on foreign energy supplies. Fifteen years ago, we were anticipating importing nearly all of our natural gas through as many as 48 import terminals. Now, we produce enough natural gas for domestic use, and we export natural gas to more than two dozen countries – our friends and allies around the world.
   a) Natural gas has created thousands of high paying jobs. If we were to replace natural gas with renewable energy would these jobs go away?
      Currently, it would be impossible to replace natural gas with renewables. If it were possible to go to a 100% renewable energy economy, the cost of just storing electricity would be 19 times the annual total electricity bill in the United States. Should some miracle technology breakthrough occur that

allows renewables to replace natural gas completely, then yes, the great paying natural gas jobs would go away.

b) What do our emissions look like compared to the rest of the world?

Unites States has achieved a 28% reduction in CO_2 emissions in the electric power sector. No more than one or two other countries can claim this accomplishment. Most other countries are seeing increases in CO_2 emissions. Some, such as China and India, are seeing dramatic increases.

c) What do you attribute to the decline in greenhouse gas emissions?

Our dramatic reduction in greenhouse emissions has occurred due to fuel switching from coal to natural gas, and significant gains in energy efficiency, including automobile efficiency, building efficiency, and gains in appliance efficiency. There has been GHG reduction from deploying renewables, but the primary drivers are fuel switching and efficiency.

In: Climate Change
Editor: Fukui Ayaka
ISBN: 978-1-53619-239-1

Chapter 2

EARTH'S THERMOMETERS: GLACIAL AND ICE SHEET MELT IN A CHANGING CLIMATE*

Committee on Science, Space, and Technology

Thursday, July 11, 2019
House of Representatives,
Committee on Science, Space, and Technology,
Washington, D.C.

The Committee met, pursuant to notice, at 10:01 a.m., in room 2318 of the Rayburn House Office Building, Hon. Eddie Bernice Johnson [Chairwoman of the Committee] presiding.

---

* This is an edited, reformatted and augmented version of Hearing Before the Committee on Science, Space, and Technology, House of Representatives, One Hundred sixteenth Congress, First Session, Serial No. 116–35, dated July 11, 2019.

Committee on Science, Space, and Technology

U.S. House of Representatives Hearing Charter:

Earth's Thermometers: Glacial and Ice Sheet Melt in a Changing Climate

Thursday, July 11, 2019

10:00 a.m.

2318 Rayburn House Office Building

PURPOSE

The purpose of this Full Committee hearing is to discuss the current state of the science on glacier and ice sheet melt. The hearing will focus on the science to understand the physical processes and projections of mass loss of the major ice sheets in Greenland and Antarctica, as well as of mountain and other land-based glaciers. The Committee will receive expert testimony on current projections of glacier mass loss due to anthropogenic climate change, and in turn how that will affect sea level. Additionally, this hearing will provide an opportunity to discuss the major sources of uncertainty related to glacial and ice sheet melt including research gaps, risks to communities from local glacier melt, as well as global risks from ice sheet instability and sea level rise, and the need for adaptation and mitigation.

WITNESSES

- Dr. Richard B. Alley, Evan Pugh Professor of Geosciences and Associate of the Earth and Environmental Systems Institute, Pennsylvania State University
- Dr. Robin E. Bell, Lamont Research Professor, Lamont-Doherty Earth Observatory, Columbia University

- Dr. Twila A. Moon, Research Scientist, National Snow and Ice Data Center's (NSIDC) Cooperative Institute for Research in Environmental Sciences
- Dr. Gabriel J. Wolkon, Research Scientist and Manager, Climate and Cryosphere Hazards Program, Division of Geological & Geophysical Surveys, Alaska Department of Natural Resources
- Dr. W. Tad Pfeffer, Fellow, Institute of Arctic and Alpine Research, University of Colorado Boulder

OVERARCHING QUESTIONS

- How is anthropogenic climate change, particularly rising temperatures, affecting glaciers and ice sheets?
- How do glaciers and ice sheets act as "Earth's thermometers" and what can past climates tell us about current rates of change?
- How much do melting glaciers and ice sheets contribute to sea level rise and what are the projections for future sea level changes?
- What methods do scientists use to study glaciers and ice sheets, and what are the major challenges and sources of uncertainty in understanding glacial and ice sheet melt?
- What is our understanding of "tipping points" or "thresholds" in ice sheet and glacial melt, such as at the Western Antarctic Ice Sheet?
- Globally, how does mountain glacial melt impact human society?

Background

Glaciers are defined as persistent, land-based, dense ice formations that form when accumulation of snow exceeds its ablation (melting and other forms of loss) over many years.[1] The world's approximately 198,000 glaciers cover less than one percent of Earth's land surface,[2] yet glacial ice is the largest reservoir of freshwater on Earth[3] and is an important source of water for plants, animals, and humans where they occur in temperate regions and release meltwater in the summer. Glaciers and ice sheets play a critical role in Earth's air and water cycles, ecosystem support through providing nutrients and shelter for plants and animals, and climate system.

Glaciers that are larger than 50,000 km^2 (20,000 mi^2) are called ice sheets, or continental glaciers.[4] The Earth's only two present day ice sheets are in Antarctica and Greenland.[5] Ninety percent of the Earth's ice mass is contained in the Antarctic ice sheet, the world's largest single ice mass, covering almost 14 million km^2 (5.4 million mi^2) and containing 30 million km^3 of ice.[6] The Greenland and Antarctic ice sheets hold enough water to raise sea levels by 65 m (over 213 ft); however, complete melting of the ice sheets is not expected to happen.[7]

Ninety-nine percent of glacial ice is contained in ice sheets in the polar regions, but mountain glaciers exist on every continent except Australia.[8] In the U.S., glaciers can be found in Washington, Oregon, California, Montana, Wyoming, Colorado, and Nevada, with the majority occurring in Alaska.[9] Glaciers are sometimes called "Earth's thermometers" because they are very sensitive to, and therefore indicators of, climatic changes.[10]

---

1 "What is a glacier?" National Snow & Ice Data Center. https://nsidc.org/cryosphere/glaciers/questions/what.html

2 Davies, B. "Mapping the world's glaciers." 2017. http://www.antarcticglaciers.org/glaciers-and-climate/glacier- recession/mapping-worlds-glaciers/

3 "Ice, Snow, and Glaciers and the Water Cycle." USGS. https://www.usgs.gov/special-topic/water-science- school/science/ice-snow-and-glaciers-and-water-cycle?qt-science_center_objects=0#qt-science_center_objects

4 "What is an ice sheet?" National Snow & Ice Data Center (NSIDC). https://nsidc.org/cryosphere/quickfacts/icesheets.html

5 Ibid.

6 Amos, Jonathan (2013-03-08). "*BBC News - Antarctic ice volume measured*". Bbc.co.uk.

7 Shepherd, A. et al. *Science*. 338: 1183-1189 (2012).

8 "State of the Cryosphere: Mountain Glaciers." NSIDC. https://nsidc.org/cryosphere/sotc/glacier_balance.html

9 USGS. "Where are glaciers found in continental North America?" https://www.usgs.gov/faqs/where-are-glaciers- found-continental-north-america?qt-news_science_products=0#qt-news_science_products

10 Moon, T. *et al.* 2018. "Rising oceans guaranteed: Arctic land ice loss and sea level rise." *Current Climate Change Reports*. https://doi.org/10.1007/s40641-018-0107-0

Glaciers are also indicators of past climates because trapped air bubbles reveal past atmospheric conditions from thousands of years ago.[11] Glaciers and ice sheets have a slow response time to global warming, and glaciers have not yet caught up to the heat additions made in the past decades. Therefore, even if global carbon emissions stopped entirely today, glaciers are locked in to a certain amount of melt. Thousands of studies conducted by researchers around the world have documented melting glaciers, diminishing snow cover, and rising sea levels.[12]

This hearing is focusing on land-based ice, as opposed to sea ice. Arctic sea ice is also diminishing at rapid rates due to anthropogenic warming,[13] but as it melts, it does not contribute to sea level changes unlike glacial and ice sheet melt.[14]

State of the Science on Glacial and Ice Sheet Melt

Globally, land-based ice is deteriorating. According to the Intergovernmental Panel on Climate Change Fifth Assessment Report (IPCC AR5), a diverse range of observational evidence from multiple data sources and independent analysis techniques provide consistent evidence of substantial retreat of mountain glaciers since the 1960s, and the increased surface melting of the Greenland ice sheet since 1993, due to warmer temperatures. According to satellite data, Greenland lost an average of 269 gigatons[15] of ice, equivalent to about 71 trillion gallons of water, per year between 2002 and 2016, with the pace accelerating in recent years (Figure 1).[16] Increased surface melt, runoff, and outlet glacier discharge from warmer air temperatures are the primary contributing factors. The portion of the Greenland Ice Sheet experiencing annual melt

[11] "Glaciers and climate change." NSIDC. https://nsidc.org/cryosphere/glaciers/ questions/ climate.html

[12] IPCC, 2014: Climate Change 2014: Synthesis Report. Contribution of Working Groups I, II and III to the Fifth Assessment Report of the Intergovernmental Panel on Climate Change [Core Writing Team, R.K. Pachauri and L.A. Meyer (eds.)]. IPCC, Geneva, Switzerland, 151 pp (IPCC AR5)

[13] "Arctic sea ice minimum." NASA. https://climate.nasa.gov/vital-signs/arctic-sea-ice/

[14] Appell, D. "Loss of land ice (not sea ice) = more sea level rise." 2014. https:// www.yaleclimateconnections.org/2014/11/loss-of-land-ice-not-sea-ice-more-sea-level-rise/

[15] 1 gigaton = 10^9 tons

[16] IPCC AR5

has increased since 1980, including through significant melting events. For example, an unprecedented 98.6% of the Greenland Ice Sheet surface experienced melt on a single day in July 2012.[17] While there are seasonal patterns of warm-weather ice melt and re-freezing in winter months, when seasonal melt outpaces re-freezing, there is net annual ice mass loss. Just last month, the National Snow and Ice Data Center (NSIDC) in Boulder, Colorado reported that the Greenland Ice Sheet appears to have experienced its biggest mid-June melt event on record.[18]

The Western Antarctic Ice Sheet (WAIS), the portion of the Antarctic ice sheet that covers the western part of the continent, is considered the most vulnerable ice sheet on Earth because its bed lies thousands of feet below sea level and is exposed to warm ocean currents.[19] In 2018, a major joint U.S.-UK research collaboration was initiated to study the possibility of "marine ice sheet instability" and "marine ice cliff instability" of the WAIS, focusing on the marine- terminating Thwaites Glacier.[20] Marine ice cliff instability is when a tall cliff that might form at the front of the glacier begins to calve and break in a runaway fashion. Marine ice sheet instability is an inherently unstable architecture caused by atmospheric and ocean warming, which could result in a positive feedback loop of rapid melting of the WAIS, triggering rapid sea level rise. The Thwaites Glacier has increased speed of movement and simultaneously experienced rapid ice thinning. Multiple studies indicate that this collapse is underway in the WAIS and may also be a cause of rapid ice front retreat occurring in Greenland.[21,22]

---

[17] Ibid.

[18] Samenow, Jason. June 14, 2018. "Temperatures leap 40 degrees above normal as the Arctic Ocean and Greenland ice sheet see record June melting." https://www.washingtonpost.com/weather/2019/06/14/arctic-ocean-greenland- ice-sheet-have-seen-record-june-ice-loss/?utm_term=.2f92ed415b9d

[19] Fox, Douglas. "The West Antarctic Ice Sheet Seems to Be Good at Collapsing." *National Geographic*, National Geographic Society, 13 June 2018, www.news.nationalgeographic.com/2018/06/west-antarctic-ice-sheet-collapse- climate-change/

[20] The International Thwaites Glacier Collaboration, https://thwaitesglacier.org/about/itgc

[21] Moon, Twila. May 2017. "Saying goodbye to glaciers: Glacier volume is shrinking worldwide, with wide-ranging implications for society." Science. Vol. 356, Issue 6338.

[22] BBC 30 April 2018. "Thwaites Glacier: Biggest ever Antarctic field campaign." By Jonathan Amos https://www.bbc.com/news/science-environment-43936372

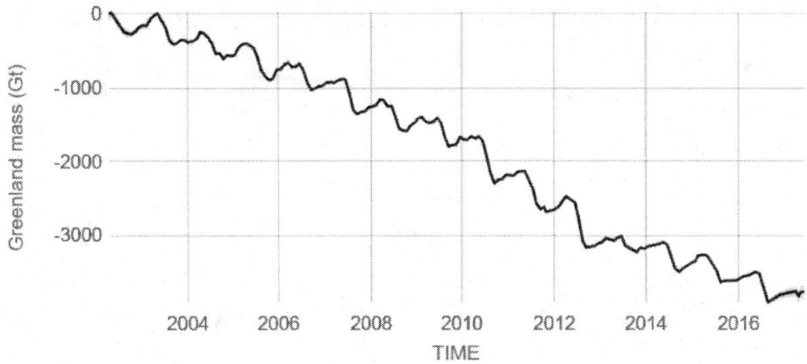

Source: climate.nasa.gov

Figure 1. Greenland ice sheet mass variation since 2002, showing loss of 286 gigatons ice per year. Data source: Ice mass measurement by NASA's Gravity Recovery and Climate Experiment (GRACE) satellites. Credit: NASA.

The WAIS has been experiencing mass loss since the early 1990s, and melt rates have more than tripled in the last 25 years.[23] Recent observed rapid mass loss from West Antarctica's floating ice shelves is attributed to increased glacial discharge rates due to diminishing ice shelves caused by the surrounding ocean becoming warmer.[24] Antarctica as a whole lost more than 3 trillion tons of ice between 1992 and 2017.[25] More recent gravity data collected from space using NASA's Gravity Recovery and Climate Experiment (GRACE) satellites show that Antarctica has been losing more than one hundred km^3 (24 mi^3) of ice each year since 2002 (Figure 2).[26] A chunk of ice the size of Delaware broke off on July 12, 2019 from the

[23] Harvey, C. "Antarctic Melt Rate Has Tripled in the Last 25 Years." June 14, 2018. https://www.scientificamerican.com/article/antarctic-melt-rate-has-tripled-in-the-last-25-years/

[24] IPCC AR5

[25] Harvey, C. "Antarctic Melt Rate Has Tripled in the Last 25 Years." June 14, 2018. https://www.scientificamerican.com/article/antarctic-melt-rate-has-tripled-in-the-last-25-years/

[26] Conway, E. "Is Antarctica melting? – Climate Change: Vital Signs of the Planet." September 16, 2014. https://climate.nasa.gov/news/242/is-antarctica-melting

Larsen C Ice Shelf of the WAIS, which might destabilize the entire ice shelf.[27]

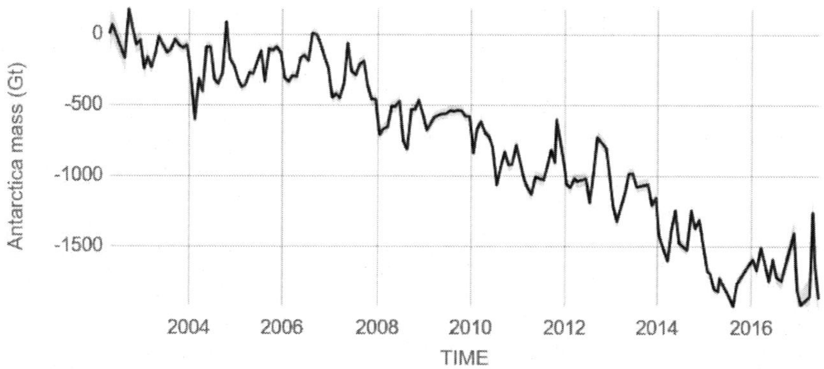

Source: climate.nasa.gov

Figure 2. Antarctic ice sheet mass variation since 2002, showing loss of 127 gigatons ice per year. Data source: Ice mass measurement by NASA's GRACE satellites. Credit: NASA.

The IPCC AR5 predicts that global glacier volume, excluding glaciers on the periphery of Antarctica and the Greenland and Antarctic ice sheets, is projected to decrease by 15 to 55% if we limit global temperature rise to 2 degrees Celsius (RCP2.6) and by 35 to 85% under a high emissions scenario (RCP8.5). However, uncertainties remain in the quantification and modeling of key physical processes that contribute to the acceleration of land ice melting. Climate models are unable to capture the rapid pace of observed land ice melt over the last 15 years; a major factor is our inability to quantify and accurately model the physical processes driving the accelerated melting.[28]

The vast majority of global mountain glaciers are losing mass at significant rates. The annually averaged ice mass from 37 global reference glaciers has decreased every year since 1984. Annual average near-surface air temperatures across Alaska and the Arctic have increased over the last

---

[27] Fox, Douglas. "The West Antarctic Ice Sheet Seems to Be Good at Collapsing." *National Geographic*, 13 June 2018, www.news.nationalgeographic.com/2018/06/west-antarctic-ice-sheet-collapse-climate-change/

[28] IPCC AR5

50 years at a rate more than twice as fast as the global average temperature. Due to increasing temperatures, Alaska is losing about 75 billion tons of ice each year.[29]

A recent study provides the clearest picture of Himalayan glacier loss to date, using declassified U.S. spy satellite data from the past 40 years and combining it with contemporary satellite data.[30] The study found that the Himalayas lost 25% of their ice in the last 40 years, equivalent to eight billion tons of water each year. Glacial melt rates tracked temperature increases during the time period studied, meaning the melting can be attributed to warmer temperatures.

Methods for Studying Ice Loss

Measuring mass loss of glaciers is not straightforward but can be estimated used many types of observations, namely, the gravitational pull of the ice, surface elevation of ice sheets, and difference between ice accumulation and loss. The modern study of mass change in glaciers and ice sheets occurs over many spatiotemporal scales, from paleo-glaciological records dating back hundreds of thousands of years, to centimeter-scale in situ measurements, to global images using satellites.[31] Field studies enable detailed sampling and long-term monitoring and involve the use of various instruments. Global Positioning System (GPS), weather stations, seismometers, time- lapse cameras, and radar instruments elucidate glacier hydrology, subsurface environments, and glacier dynamics on time scales of minutes to months.[32] Challenges in observational work are that it is time-consuming, involves travel to remote and difficult to access locations, is expensive, and involves physically demanding work in harsh environments.

---

[29] Fountain, H. "When the Glaciers Disappear, Those Species Will Go Extinct." April 17, 2019. https://www.nytimes.com/interactive/2019/04/16/climate/glaciers-melting-alaska-washington.html

[30] Maurer, J.M. et al. 2019. "Acceleration of ice loss across the Himalayas over the past 40 years." Science Advances. Vol 5, no. 6. DOI: 10.1126/sciadv.aav7266

[31] Moon, Twila. May 2017. "Saying goodbye to glaciers: Glacier volume is shrinking worldwide, with wide-ranging implications for society." Science. Vol. 356, Issue 6338.

[32] Moon, T. *et al.* 2018. "Rising oceans guaranteed: Arctic land ice loss and sea level rise." *Current Climate Change Reports.*

The advent of satellite monitoring in the 1990s was a major advancement for studying large ice sheets, allowing for improvements in the ability to estimate ice mass loss. NASA's GRACE satellite mission (2002-2017) and GRACE Follow On (launched May 2018)[33] help estimate ice mass variations, altimetry satellites (e.g., NASA's Ice, Cloud, and land Elevation Satellite, ICESat (2003-2009) and ICESat-2 (launched 2018)[34] and the European CryoSat (1999-2005) and CryoSat-2 (launched 2010))[35] detect changing surface elevations, and optical and radar imaging satellites measure ice motion, monitor glacier advance and retreat, and observe surface properties, including melt. Aerial surveys are able to cover inaccessible regions such as glacial crevasses and help with data collection in between satellite missions.[36]

Observations of ice sheets and glaciers and past records of glaciation are important for understanding Earth's climate system. Observational data is important in validating models in order to predict future changes. Ice sheets are important to the climate system, and incorporating ice sheet models into global climate models will improve projections. However, this is a challenge given ice sheet models are very high resolution and the climate models cannot currently accommodate that level of detail.[37]

Resulting Sea Level Rise

Sea levels have risen over eight inches (23 cm) since the Industrial Revolution and continue to rise 0.13 inches (3.2 mm) each year.[38] One third of current sea level rise is due to thermal expansion of seawater, one third is from ice sheet melt, and one third is from mountain glacial melt. The contribution to sea level rise from Antarctic and Greenland ice sheet melt has gone up from one-tenth just two decades ago, mostly due to Greenland

---

[33] The NASA GRACE mission concluded in June 2017 and GRACE's successor mission, GRACE Follow-On, is launching in the summer of 2019.

[34] NASA ICESat and ICESat-2, https://icesat.gsfc.nasa.gov/

[35] European Satellite Agency, CryoSat and CryoSat-2, https://www.esa.int/Our_Activities/Observing_the_Earth/CryoSat/Introducing_CryoSat

[36] Moon, T. *et al.* 2018. "Rising oceans guaranteed: Arctic land ice loss and sea level rise." *Current Climate Change Reports.*

[37] Fourth National Climate Assessment.

[38] Nunez, C. 2017. "Sea level rise, explained." https://www.nationalgeographic.com/environment/global- warming/sea-level-rise/

ice losses.[39] The Fourth National Climate Assessment (NCA4) predicts global mean sea level will rise an additional 1.0-4.3 feet by 2100, depending on a low or high emissions scenario.[40] Even with a low emissions scenario and not considering additional contribution from ice sheet melt, we are locked in to approximately 1 foot of global sea level rise by the end of the century.[41]

Over the last two decades, significant progress has been made in understanding ice sheet dynamics through combined field and satellite observations and improving numerical models to capture responses of ice sheets to environmental change. However, there is major uncertainty in the amount of additional sea level rise that could occur due to melting of Greenland and Antarctic ice sheets, due to the possibility of marine ice sheet and marine ice cliff instability. Collapse of the WAIS could contribute an additional 11 feet of sea level rise.[42] Sea level rise threatens coastal communities in the U.S. and worldwide and will increase the frequency and extent of extreme flooding associated with coastal storms, such as hurricanes and nor'easters.[43]

Other Impacts of Glacial and Ice Sheet Melt

Glacial and ice sheet mass loss will also have other direct and indirect impacts on humans and ecosystems, including on the climate system and weather patterns, ocean circulation, Earth's rotation, drinking water, and certain fisheries. Mountain glaciers are an important drinking water supply for many people around the world, especially in India, Nepal, and some countries in South America. Approximately 800 million people depend on glacial meltwater from the high mountains of Asia alone.[44] Glacial melt

---

[39] Shepherd, A. et al. Science. 338: 1183-1189 (2012).

[40] NCA4; Volume I; Ch. 12

[41] Ibid

[42] Fox, Douglas. "The West Antarctic Ice Sheet Seems to Be Good at Collapsing." *National Geographic*, National Geographic Society, 13 June 2018, www.news.nationalgeographic.com/2018/06/west-antarctic-ice-sheet-collapse- climate-change/

[43] IPCC AR5

[44] Pritchard, H.D. 2019. "Asia's shrinking glaciers protect large populations from drought stress." Nature, Vol 569. https://doi.org/10.1038/s41586-019-1240-1

leads to rapid, catastrophic floods and debris flows for these downstream communities.[45]

Recent research suggests ice sheet melt may substantially slow down the major ocean conveyor belt of heat, known as the Atlantic meridional overturning circulation, which helps regulate the climate and affects global weather patterns.[46] There is satellite evidence that ice sheet melt is also responsible for a slight decrease in the speed of the Earth's rotation. This is because ice sheets are at high latitudes, and when they melt, the water is redistributed toward lower latitudes in a phenomenon called "polar wander."[47]

Loss of glacial streams and meltwater will lead to extinction of small creatures that rely on them. If glacial meltwater continues to decline and stream temperatures rise, larger fish populations like salmon and similar fish may also be affected.[48]

Chairwoman JOHNSON. Good morning. This hearing will come to order. And without objection, the Chair is authorized to declare recess at any time.

I'd like to welcome our witnesses to the Science, Space, and Technology Committee's hearing entitled, "Earth's Thermometers: Glacial and Ice Sheet Melt in a Changing Climate." It seems as though we're bombarded on an almost daily basis with news articles and reports saying that the world's ice is melting faster than ever. As a matter of fact, I almost invited Mr. Young from Alaska, who moved to Alaska because it was too warm in the United States proper. Since I read about Alaska last week, I thought he might want to hear this.

Pictures show ice sheets in Greenland and Antarctica crashing into the oceans before our eyes. Just last month, a piece of ice the size of the State of Delaware broke off Antarctica, and Greenland was reported to have

---

[45] Ibid

[46] Harvey, C. 2019. "Melting ice sheets could worsen extreme weather." https://www.scientificamerican.com/article/melting-ice-sheets-could-worsen-extreme-weather/

[47] Dunham, W. 2015. "Melting glaciers blamed for subtle slowing of Earth's rotation." https://www.reuters.com/article/us-science-rotation-idUSKBN0TU2F720151212

[48] Ibid

experienced the biggest June ice melt event on record with temperatures 40 degrees above normal.

The rate of change in the Arctic and Antarctic has been quickening in recent years, according to the Intergovernmental Panel on Climate Change (IPCC) and numerous other scientific bodies. For example, a study published in Nature in January that was led by an international team of more than six dozen researchers tells us that melt rates have more than tripled in western Antarctica in the last 25 years.

Mountain glaciers are also experiencing rapid rates of change. Just a few weeks ago, declassified U.S. spy satellite data clearly showed that Himalayan glaciers lost 25 percent of their ice over the last 40 years. This is equivalent to 8 billion tons of water each year. This puts the hundreds of millions of people in that region who depend on glacial melt as a freshwater source at risk.

According to the 2014 IPCC Assessment Report, without significant reductions in global greenhouse gas emissions, mountain glaciers will lose 35 to 85 percent of their ice by the end of the century under a high emissions scenario. Newer reports indicate that IPCC estimates might even be conservative and that glacial and ice sheet melt rates could even be higher. We need to be listening to Earth's glaciers and ice sheets and what they're telling us about the changing climate.

Glacial and ice sheet melt is responsible for two-thirds of the 8 inches of sea-level rise that we've seen in the last 200 years from the anthropogenic warming, and that sea-level rise is only expected to continue. The western Antarctic ice sheet, which everyone is watching because it is thought to be the most unstable ice sheet, could add another 11 feet of additional sea-level rise if it collapses, which some experts expect could happen at some point. Such an increase would mean many coastal cities would be flooded, and the world as we know it would be different.

What's happening in Greenland, Antarctica, and the high mountain regions matters to us all. Glaciers and ice sheets play vital roles in regulating Earth's climate and weather, provide over two-thirds of the Earth's freshwater supply for drinking and agricultural uses, support fisheries and ecosystem health, and run hydro-power plants. I'm glad we have the opportunity to hear today from some of the Nation's leading glacial and ice sheet experts.

And I'd like to welcome Dr. Richard Alley, who last testified before this Committee in 2010. I also want to announce that later today we will be hosting a screening of the award-winning documentary "Chasing Ice" that documents changing ice in the Arctic. It will be followed by a question-and-answer session with two of our witnesses, Dr. Pfeffer, who was a scientific advisor to the film, and Dr. Moon. The screening is free and open to the public, and I hope all of you will join us.

This Committee plays an important role in authorizing both climate science and the research needed to better understand glaciers and ice sheets. Since the 1990s, NASA's (National Aeronautics and Space Administration's) ice-monitoring satellites have led to major discoveries of ice sheet dynamics and melt, while the National Science Foundation (NSF) has funded major field expeditions in ice sheets. I look forward to today's discussion with our distinguished panel to understand how Congress and the Committee in particular can address the critical research gaps in this field.

Thank you.

[The prepared statement of Chairwoman Johnson follows:]

Good morning. I would like to welcome our witnesses to the Science, Space, and Technology Committee's hearing entitled "Earth's Thermometers: Glacial and Ice Sheet Melt in a Changing Climate."

It seems as though we're bombarded on an almost daily basis with news articles and reports saying that the world's ice is melting faster than ever. Pictures show ice sheets in Greenland and Antarctica crashing into the oceans before our eyes. Just last month, a piece of ice the size of Delaware broke off of Antarctica, and Greenland was reported to have experienced the biggest June ice melt event on record with temperatures *40 degrees above normal.*

The rate of change in the Arctic and Antarctic has been quickening in recent years, according to the Intergovernmental Panel on Climate Change and numerous other scientific bodies. For example, a study published in

Nature in January that was led by an international team of more than six dozen researchers tells us that melt rates have more than tripled in Western Antarctica in the last 25 years.

Mountain glaciers are also experiencing rapid rates of change. Just a few weeks ago, declassified U.S. spy satellite data clearly showed that Himalayan glaciers lost 25% of their ice over the last 40 years. That is equivalent to eight billion tons of water each year. This puts the hundreds of millions of people in that region who depend on glacial melt as a fresh water source at risk.

According to the 2014 IPCC Assessment Report, without significant reductions in global greenhouse gas emissions, mountain glaciers will lose 35 to 85% of their ice by the end of the century under a high emissions scenario. Newer reports indicate that the IPCC estimates might even be conservative and that glacial and ice sheet melt rates could be even higher.

We need to be listening to Earth's glaciers and ice sheets and what they're telling us about the changing climate. Glacial and ice sheet melt is responsible for two-thirds of the 8 inches of sea level rise we've seen in the last 200 years from anthropogenic warming, and that sea level rise is only expected to continue. The Western Antarctic Ice Sheet, which everyone is watching because it is thought to be the most unstable ice sheet, could add another 11 feet of additional sea level rise if it collapses, which some experts expect could happen at some point. Such an increase would mean many coastal cities would be flooded and the world as we know it would be different.

What's happening in Greenland, Antarctica, and in high mountain regions matters to us all. Glaciers and ice sheets play vital roles in regulating Earth's climate and weather, provide over two-thirds of Earth's freshwater supply for drinking and agricultural uses, support fisheries and ecosystem health, and run hydropower plants. I'm glad we have the opportunity to hear today from some of the nation's leading glacial and ice sheet experts. We're lucky to have five distinguished glaciologists here today, and I would like to welcome back Dr. Richard Alley, who last testified before this Committee in 2010.

I also want to announce that later today we will be hosting a screening of the award-winning documentary *Chasing Ice* that documents changing ice in the Arctic. It will be followed by a question and answer session with two of our witnesses, Dr. Pfeffer (FEFF-er), who was a scientific advisor to the film, and Dr. Moon. The screening is free and open to the public, and I hope you can join us.

This Committee plays an important role in authorizing both climate science and the research needed to better understand glaciers and ice sheets. Since the 1990s, NASA's ice monitoring satellites have led to major discoveries of ice sheet dynamics and melt, while the National Science Foundation has funded major field expeditions to ice sheets. I look forward to today's discussion with our distinguished panel to understand how Congress, and this Committee in particular, can address the critical research gaps in this field. Thank you.

Chairwoman JOHNSON. And I now will offer our Ranking Member his opening statement time.

Mr. LUCAS. Thank you, Chairwoman Johnson, for holding this hearing, which is another opportunity to examine the impacts of a changing climate on our country and the world at large. While today's hearing will examine the underlying science of this issue and concerns about climate change, I'd like for us to also focus on the agricultural, economic, and geopolitical consequences we can expect from glacial and sea ice melt and, more importantly, how we can address those.

For instance, polar ice sheets cool ocean currents, which affect global weather patterns. As I've mentioned a time or two, weather issues are of paramount importance to farmers and ranchers in Oklahoma and around the world. We do not have a firm grip on how these weather patterns will change due to melting and how we can prepare for these changes.

I also want to consider the economic and geopolitical consequences of glacial and sea ice melt. Five countries, including America and Russia, border the Arctic. Territorial disputes in this region will take on greater importance as resource-rich land and new shipping routes are revealed.

There are significant economic implications from the energy rights, mineral deposits, and tourism opportunities. For instance, Russia is claiming that some newly accessible routes should not be considered international waterways but a part of their sovereign territory. Better research will give us greater insights into how we can expect shipping routes to change so we can prepare to address these issues.

As the Science Committee, we have a responsibility to address our national research priorities, and those must be broader than just how the climate's changing. We need to understand the specific effects so we can adopt and continue our economic growth.

During our first full hearing of this Congress, Members of the Committee discussed how we could embrace a broader portfolio of basic research, energy innovation, and competitive technology to make energy production cleaner, more efficient, and less costly. I hope we can spend more time considering research into innovative technologies like nuclear reactors, battery storage, and carbon capture.

I'd like to thank our witnesses for being here today, and I look forward to our discussion. And I yield back, Madam Chair.

[The prepared statement of Mr. Lucas follows:]

Chairwoman Johnson, thank you for holding this hearing, which is another opportunity to examine the impacts of a changing climate on our country and the world at large.

While today's hearing will examine the underlying science of this issue and concerns about climate change, I'd like for us to also focus on the agricultural, economic, and geopolitical consequences we can expect from glacial and sea ice melt and—more importantly—how we can address those.

For instance, polar ice sheets cool ocean currents which affect global weather patterns. As I've mentioned once or twice, weather issues are of paramount importance to farmers and ranchers in Oklahoma and around the world. We do not have a firm grasp of how these weather patterns will change due to melting and how we can prepare for these changes.

I also want to consider the economic and geopolitical consequences of glacial and sea ice melt. Five countries, including America and Russia, border the Arctic. Territorial disputes in this region will take on greater importance as resource-rich land and new shipping routes are revealed.

There are significant economic implications from the energy rights, mineral deposits, and tourism opportunities. For instance, Russia is claiming that some newly accessible routes should not be considered international waterways but part of their sovereign territory. Better research will give us greater insight into how we can expect shipping routes to change so we can prepare to address these issues.

As the Science Committee, we have the responsibility to address our national research priorities and those must be broader than just how the climate is changing. We need to understand its specific effects so we can adapt and continue our economic growth.

During our first full committee hearing of this Congress, members of this Committee discussed how we must embrace a broad portfolio of basic research, energy innovation, and competitive technology to make energy production cleaner, more efficient, and less costly.

I hope we can spend more time considering research into innovative technologies like nuclear reactors, battery storage and carbon capture.

I'd like to thank our witnesses for being here today, and I look forward to our discussion.

Chairwoman JOHNSON. Thank you very much.

I'd like to extend a warm welcome to a guest in the audience, Maria, from Chandler, Arizona. Could you stand? We hear you're a rising senior in high school who's interested in studying engineering in college. And it's great to have the next generation of STEM (science, technology, engineering, and mathematics) professionals represented here today. And welcome to all the young people over here, too. Thank you for being here.

At this time I'd like to introduce our witnesses. Our first distinguished witness, Dr. Richard Alley, is the Evan Pugh Professor of Geosciences and Associate of the Earth and Environmental Systems Institute at the Pennsylvania State University. He has spent more than 40 years studying the great ice sheets to help predict future changes in climate and sea levels, and has made four trips to Antarctica, nine to Greenland, and additional expeditions to Alaska and elsewhere. He has authored or co-authored more than 300 scientific papers. He was involved in the IPCC group of contributors that won the 2007 Nobel Peace Prize. He won Pennsylvania

State's highest teaching award, and has written a book on climate change and ice cores. He holds a Ph.D. in geology from the University of Wisconsin.

Our second witness, Dr. Robin Bell, is the PGI Lamont Research Professor at Lamont-Doherty Earth Observatory of Columbia University and a member of the faculty at Columbia Earth Institute. She directs programs in ice sheet dynamics, leads efforts to develop innovative technology, and works to improve the scientific culture, especially for women. She has led 10 major expeditions to the polar regions discovering an active volcano, large, deep lakes, and hidden mountain ranges buried by ice. She was instrumental in launching the International Polar Year in 2007 that brought together over 50,000 scientists. Currently, she is the President of the American Geophysical Union, the largest collection of Earth and space scientists in the world. And her Ph.D. is in geophysics from Columbia University.

Our third witness is Dr. Twila Moon, who is a Research Scientist at the National Snow and Ice Data Center (NSIDC), part of the University of Colorado's Boulder Cooperative Institute for Research in Environmental Sciences. She studies modern changes in glaciers and ice sheets and the connection among ice, climate, ocean, and ecosystems. Her research focuses on the Greenland ice sheet and the Arctic and uses a variety of tools, including satellite remote sensing, fieldwork, and computer simulations. She also leads efforts to improve science and knowledge coproduction between scientists and stakeholders. Dr. Moon received her Ph.D. in Earth and space sciences from the University of Washington.

Our fourth witness, Dr. Gabriel Wolken, is a Research Scientist and Manager of the Climate and Cryosphere Hazards Program at the Alaska Division of Geological & Geophysical Surveys and a Research Assistant Professor at the International Research Center at the University of Alaska Fairbanks. There, he is a Senior Scientist in the Climate Adaptation Science Center. He studies snow and glacier change and their connection to climate and natural hazards through observations, remote sensing, and computer modeling. Dr. Wolken has a Ph.D. in Earth and atmospheric sciences from the University of Alberta.

Our final witness, Dr. William Ted Pfeffer, is a Professor of Civil, Environmental, and Architectural Engineering and a Fellow at the Institute of Arctic and Alpine Research at the University of Colorado Boulder. He has been involved in glaciology research for 40 years, studying the world's mountain glaciers. He has conducted hundreds of field expeditions in the continental USA, Alaska, Canada, Norway, Greenland, Antarctica, the Himalayas, and Africa. He has published over 60 peer-reviewed scientific papers and was a scientific advisor to the Emmy-winning film "Chasing Ice." Dr. Pfeffer earned his Ph.D. in geophysics at the University of Washington.

As our witnesses should know, you will each have 5 minutes for your spoken testimony. Your written testimony will be included in the record of the hearing. When all of you have completed your spoken testimony, we will begin a round of questions. Each Member will have 5 minutes to question the panel. And so we will begin our witnesses now with Dr. Alley.

TESTIMONY OF DR. RICHARD B. ALLEY, EVAN PUGH PROFESSOR OF GEOSCIENCES AND ASSOCIATE OF THE EARTH AND ENVIRONMENTAL SYSTEMS INSTITUTE, PENNSYLVANIA STATE UNIVERSITY

Dr. ALLEY. Thank you, Madam Chairwoman, Ranking Member Lucas, distinguished Members, staff, and citizens, for this opportunity to address you.

We have high scientific confidence that the world is warming primarily because we burn fossil fuels and release CO2, and this is having broad-based impacts. You've asked us to tell you about changes in snow and ice of which we will get to some of them but not all.

We still have winter, we still have blizzards. Where and when snow and ice care about temperature we are seeing broad-based shrinkage, and this really is having impacts. Earlier spring snow melt means that you can lengthen the fire season. It affects eco-systems; it affects tourism. Loss of Arctic sea ice, as Representative Lucas mentioned, has national security implications, as well as weather implications. Glacier melt is changing streamflow in some of the most overused and politically sensitive rivers on Earth.

I will focus particularly on sea level, which is the biggest global footprint of melting ice. Sea level is rising. Recently, it's been about 1 inch per 8 years. It is rising not because of natural cycles but because of warming. The ocean expands as it warms. The mountain glaciers are melting. The edges of Greenland are melting and putting extra water into the ocean. And there's faster flow of non-floating ice into the ocean from parts of Greenland and Antarctica.

We are committed to some additional sea-level rise. Just as if you drop an ice cube into your tea, it is committed to melting, but it takes a while to melt. The ice has not caught up with the warming we have already caused. But by the time our students are getting old, the decisions that we humans make now and in the future will grow to be the dominant control on how much sea-level rise we experience.

This sea-level rise is already having implications. You can Google the picture of the octopus in the parking garage in Miami on a high tide, not a storm. But the impacts could become much larger. The general projections are that if we don't change our energy system, we will get something like 3 feet of sea-level rise by 2100 above the natural level, the pre-industrial level.

And I'd like to speak about the uncertainties in that, right? So I'd like to do an analogy first. I ride my bicycle to work at Penn State. My wife drives our car. But I drove down here. I saw commuters in the D.C. area. My impression is that a commuter in D.C. expects to spend half an hour stuck in traffic. The best thing that can happen to a commuter is no traffic, but they might spend an hour, and they might get run over by a drunk driver and be in the hospital or worse. What they expect, the most likely future, is well on the good end of the possible futures when you get in that car.

When we look at the sea-level rise, it is similar. Three feet if we don't change our energy system, maybe 2, maybe 4, maybe 5, 10. We're not sure. It could be much worse. And there isn't much better to offset the much worse. There are drunk drivers in the climate system.

I'd like to explain one of them. If you ever get the chance to go to Glacier Bay National Park and Preserve in Alaska, it is a gloriously beautiful place. You can cruise 65 miles up the Bay and see little glaciers breaking off little icebergs in shallow water, and it's still spectacular. When Vancouver was on his cruise in 1796, there was no Glacier Bay. It was entirely full of ice up to a mile thick. When John Muir went by, less than a century later, the Bay was mostly open because icebergs had been breaking off the front of the glacier like dominoes at a rate of up to 7 miles a year, falling over.

That process has happened to other glaciers in Alaska. You have world experts on that process here. It has happened in Chile, in Svalbard. It's happening in Greenland and the Antarctic Peninsula.

It happened to ice sheets in the past. And it's well-known that this happens when it gets too warm where ice flows into the ocean. So far, those have been in narrow valleys. They're spectacular locally but one collapse doesn't raise global sea level a lot. If this starts to happen in parts of Antarctica rather than a narrow valley, it will open into a broad embayment. If that breaks as rapidly as we have seen elsewhere, in the next century you might get 10 feet or so of extra sea-level rise. It could be faster than that.

It is very clear that the uncertainties can be reduced if you fund bright young people to work with the co-panelists up here. That's self-serving, but it's correct. But there may be a little irreducible uncertainty in the same way that you can never predict where every drunk driver might be out on the highway. If we raise temperature, we raise sea level with high confidence, and the uncertainties are it could be a little better, a little worse, or a lot worse. Thank you.

[The prepared statement of Dr. Alley follows:]

Antarctic Contributions to Future Sea-Level Rise: Possible Tipping Points. Testimony of Dr Richard B. Alley*, Pennsylvania State University.

For the hearing entitled Earth's Thermometers: Glacial and Ice Sheet Melt in a Changing Climate Before the Committee on Science, Space and Technology, United States House of Representatives, 10:00 a.m., July 11, 2019

Personal Introduction and Background

My name is Richard Alley. I am an Evan Pugh University Professor of Geosciences and Associate of the Earth and Environmental Systems Institute at the Pennsylvania State University. I have authored or coauthored more than 300 refereed scientific papers, and I have made more than 1000 public presentations concerning my areas of expertise. I have been recognized with awards for research, teaching and service, including election to the US National Academy of Sciences and foreign membership in the Royal Society; as noted above, my comments are my own and do not represent these bodies or any other bodies, but the recognition by these bodies may help establish my credentials as a witness.

My research is especially focused on the great ice sheets of Antarctica and Greenland, their potential for causing major changes in sea level, the climate records they contain, and their other interactions with the environment; I also study mountain glaciers, and ice sheets of the past. I have served with distinguished national and international teams on major scientific assessment bodies, including chairing the U.S. National Research Council's Panel on Abrupt Climate Change (report published in 2002), and serving the U.S. Climate Change Science Program, and the Nobel-Peace-Prize-Winning Intergovernmental Panel on Climate Change (IPCC) in various ways on their Second (1995), Third (2001) and especially Fourth

* Any opinions, findings, conclusions, or recommendations expressed in this publication are those of the author and do not necessarily reflect those of the Pennsylvania State University, the Intergovernmental Panel on Climate Change, the National Academy of Sciences, or other organizations.

(2007) Assessment Reports. I have had the honor on several occasions of providing requested testimony and briefings to high government officials at the federal as well as state levels, including to legislative committees chaired by members of both major political parties, and to executive officials in administrations of both major political parties, drawing on my expertise to provide scientific information to those working for the public good. Additional information is given in the short biography at the end of this document.

My testimony here is updated from my testimony of November 17, 2010 to the Subcommittee on Energy and Environment of the House Committee on Science and Technology of the United States House of Representatives; the consistency of this testimony reflects the consistency of the scientific understanding, which continues to strengthen without fundamentally changing.

Background on Climate Change, Global Warming and Sea-Level Rise

Scientific assessments such as those of the National Academy of Sciences of the United States (e.g., National Research Council, 1975; 1979; 2001; 2006; 2008; 2010a; 2010b), the U.S. Climate Change Science Program (CCSP), and the Intergovernmental Panel on Climate Change (IPCC) have for decades consistently found with increasingly high scientific confidence that human activities are raising the concentration of carbon dioxide and other greenhouse gases in the atmosphere, that this has a warming effect on the climate, that the climate is warming as expected, and that the changes so far are small compared to those projected if humans burn much of the fossil fuel on the planet.

The basis for expecting and understanding warming from carbon dioxide is the fundamental physics of how energy interacts with gases in the atmosphere. This knowledge has been available for over a century, was greatly refined by military research after World War II, and is directly confirmed by satellite measurements and other data (e.g., American Institute of Physics, 2008; Harries et al., 2001; Griggs and Harries, 2007).

With very high scientific confidence, this warming is causing sea level to rise. Since 1993, when high-quality satellite altimetry data have been available, sea-level rise has averaged approximately 3.1 mm/yr, or about 1 inch per 8 years. (Of many sources for this general information on sea-level change, Church et al., 2013 and Lemke et al., 2007 are good starting points, and important information is available at http://sealevel.colorado.edu/ and in Nerem et al., 2018.) The rate of sea-level rise has accelerated. The rise comes from several sources. Much of the energy added to the Earth system because of the rise in atmospheric greenhouse gases has heated the ocean, causing the water to expand and raise sea level. Warming is causing melting of glaciers, shifting water from land to the ocean, with local impacts on water availability and hazards as well as global implications through sea level. Melting has increased mass loss from the surface of the Greenland ice sheet. Faster flow of non-floating ice into the ocean to float and melt has caused mass loss from some parts of the Greenland and Antarctic ice sheets. In addition, groundwater "mining"—removing well-water from the ground more rapidly than replenished—may be contributing a little to sea-level rise. (Note that the IPCC found with high confidence that warming is causing loss of other temperature- sensitive snow and ice, including springtime snow cover and Arctic sea ice, but that these do not contribute in any significant way to sea-level rise; e.g., Lemke et al., 2007.)

Many strands of mutually supporting evidence are woven into the confident knowledge that loss of land ice and warming of the ocean are driving sea-level rise. A large and consistent scientific literature exists on this topic (e.g., The IMBIE Team, 2018), and the synopsis of techniques in Lemke et al. (2007; section 4.6.2.1) provides a broad overview. As explained there, for large ice sheets the gain or loss of mass contributing to sea-level change is measured in three ways: by repeatedly "weighing" the ice sheet using satellites that measure Earth's gravity field, supplemented by data from aircraft and surface measurements; by repeatedly measuring the surface elevation of the ice sheet using altimeters on satellites or aircraft supplemented by surface measurements; and, by measuring and modeling the addition of snowfall to the ice sheet, and the loss by runoff of meltwater or by flow of nonfloating ice into the ocean to float and then melt. Measurements of ocean temperature provide data to estimate the expansion of ocean water, and measurements of changes in mountain glaciers similar to those for ice sheets but involving more on-the-glacier work and more analysis of size changes from satellite imagery provide additional constraints. Sea-level measurements from satellites and tide gauges monitor the rise of the ocean. Scientists then assess the consistency among the various lines of evidence, and find agreement. Additional tests are also applied. Melting of ice from near the poles and motion of the water into the ocean actually slows Earth's rotation very slightly, something like a spinning ice skater sticking out her arms. And, while Greenland is far north, it is not all the way to the North Pole, so melting of Greenland's ice causes Earth to wobble a bit, something like the skater moving one arm but not the other. These changes in Earth's rotation are tiny, they are not important to most people most of the time, but they test and confirm the other measurements.

Insights from Climate History

The hills around Los Angeles have always burned, so we worry about people shooting illegal fireworks during dry summers. People have always died, so we worry about murder. An arson investigator must understand natural fires, and a homicide investigator must understand how people live and die naturally.

The science of climate and sea level includes a "CSI" component (see, for example, CCSP, 2009; Masson-Delmotte et al., 2013). Summarizing those summaries and many other studies, climate has always changed, which shows that climate is changeable. Climate has changed for many reasons—dust from single large volcanic eruptions blocking the sun and causing occasional cold years, features of Earth's orbit slowly shifting sunshine around over tens of thousands of years, small changes in the brightness of the sun, shifts in ocean circulation, the very slow drifting of continents—but naturally caused changes in greenhouse gases and especially carbon dioxide have been very important over Earth's climate history. When warming has occurred, ice has melted and sea level has risen. Careful study of these natural changes and their causes contributes to the strong knowledge that the changes now occurring are not primarily the continuation of some natural cycle, but are instead caused primarily by the human-driven rise in greenhouse gases, mostly carbon dioxide.

Past warming and cooling in response to changing greenhouse-gas concentrations and other forcings including features of Earth's orbit have caused much larger past changes in ice and sea level than have occurred in the last century or that are projected for the next century. But, human burning of fossil fuels could cause a climate change that rivals or exceeds those of the past in combined size and speed (White et al., 2010), raising questions about the future that are discussed next.

Looking to the Future

Some additional sea-level rise is already committed, under the future emissions pathways considered by the IPCC. Just as an ice cube placed in a glass of tea takes a while to melt, the warming of the ocean and mass loss from land ice have not yet "caught up" with the warming that has already been caused, and would continue to contribute to some additional sea-level rise even if temperature were stabilized today. Looking toward the time when today's students are old, toward the end of this century and beyond, future human decisions become increasingly important and then dominant in controlling projected sea-level rise (e.g., Church et al., 2013, Fig. 13.27). Under strong warming (RCP8.5), projected sea- level rise in 2100 compared to preindustrial is slightly more than 3 feet, with an uncertainty that the IPCC considered likely to be no more than about 1 foot (details and more-precise numbers available in the source).

One occasionally hears the unsubstantiated claim that the IPCC projections are overly alarmist. In fact, several studies have suggested that, if the IPCC is open to criticism on this subject, it is overly conservative. Rahmstorf et al. (2007), for example, compared observed sea-level rise at the time to projections from earlier IPCC work, and found that the sea was rising faster than the most-likely projection and near the projected likely upper limit. The Third Assessment Report of the IPCC projected a most-likely future in which climate change would cause growth of the Antarctic ice sheet over this century (Church et al., 2001), but Antarctic ice is shrinking (The IMBIE Team, 2018); more-recent central projections from the IPCC indicate less ice loss from Antarctica than observed (Slater and Shepherd, 2018). Garner et al. (2018) compared the history of IPCC projections to other science-based projections, many of which were available to the IPCC assessment teams; Figure 2a in Garner et al. (2018) shows no tendency for the IPCC projections to be higher than in the underlying literature, and some tendency for the IPCC projections to be lower. The new expert elicitation from Bamber et al. (2019) includes information obtained since the IPCC Fifth Assessment Report (Church et al., 2013), and yields potential for higher sea-level rise than generally projected. Quoting from the Abstract, with English units added:

> "For a +2°C (+3.6 °F) temperature scenario consistent with the Paris Agreement, we obtain a median estimate of a 26 cm (0.85 ft) SLR contribution by 2100, with a 95th percentile value of 81 cm (2.7 ft). For a +5°C (+9°F) temperature scenario more consistent with unchecked emissions growth, the corresponding values are 51 and 178 cm (1.7 and 5.8 ft), respectively. Inclusion of thermal expansion and glacier contributions results in a global total SLR estimate that exceeds 2 m (6.6 ft) at the 95th percentile. Our findings support the use of scenarios of 21st century global total SLR exceeding 2 m (6.6 ft) for planning purposes. Beyond 2100, uncertainty and projected SLR increase rapidly. The 95th percentile ice sheet contribution by 2200, for the +5°C (+9°F) scenario, is 7.5 m (24.6 ft) as a result of instabilities coming into play in both West and East Antarctica. Introducing process correlations and tail dependences increases estimates by roughly 15%."

Focus on Ice-Sheet Changes

The large ice sheets of Greenland and Antarctica are of special interest, because they are so big and thus could affect sea level so much. Expansion of the ocean as it warms gives just under 1 foot of rise per degree F of warming (0.4 m/1°C) (Levermann et al., 2013), but the ~1000-year time for heat to mix into the ocean makes the resulting sea-level rise relatively slow. Melting of all of the world's mountain glaciers and small ice caps might raise sea level by about 1 foot (0.3 m), but melting of the great ice sheets would raise sea level by just over 200 feet (more than 60 m), with ~23 feet from Greenland (7.3 m) and the rest from Antarctica (Lemke et al., 2007). We do not expect to see melting of most of that ice over the next century or centuries, but even a relatively small change in the ice sheets could matter to the world's coasts; roughly 10% of the world's population lives within 33 feet (10 m) of sea level (McGranahan et al., 2007). I thus next consider the possibility of rapid changes (National Research Council, 2002; 2013; CCSP, 2008; 2009).

Ice-Sheet Behavior

An ice-sheet is a two-mile-thick, continent-wide pile of snow that has been squeezed to ice under the weight of more snowfall. (For a more-detailed background, see Cuffey and Paterson, 2010.) Growth or shrinkage of an ice sheet depends on the balance between snowfall and melting on top, and on flow taking accumulated ice to lower elevation, often to the ocean to make icebergs.

The balance between snowfall and melting is of great importance. I will not treat this aspect in detail, however; additional information is available from the IPCC reports cited herein. Generally, warming of subfreezing conditions is expected to increase snowfall because warmer air can carry more water, although additional processes are important. For Greenland and mountain glaciers, the increase in melting from warming is expected to exceed the increase in snowfall. For Antarctica, fairly large warming would be required for mass loss by surface melting to become important, so the surface is expected to gain mass with moderate warming; however, flow increases have recently exceeded changing snowfall leading to mass loss, and this is expected to continue. I thus focus next on the flow characteristics. A review of much of this information is available in Alley et al. (2015) and Scambos et al. (2017) as well as in the IPCC reports and in Cuffey and Paterson (2010). The literature is large and increasing rapidly.

All piles, including glaciers and ice sheets, tend to spread under their own weight, restrained by their own strength (which is why spilled coffee spreads on a table top but the stronger table beneath does not spread), by friction beneath (so pancake batter spreads faster on a greased griddle than on a dry waffle iron), or by "buttressing" from the sides (so an appropriately placed spatula will slow the spreading of the pancake batter).

Some early gothic cathedrals suffered from the "spreading-pile" problem, in which the sides tended to bulge out while the roof sagged down, with potentially unpleasant consequences. The beautiful solution was the flying buttress, which transfers some of the spreading tendency to the strong earth beyond the cathedral. Ice sheets also have "flying buttresses", called ice shelves. In the coldest regions, the ice reaching the ocean usually does not immediately break off to form icebergs, but remains attached to the ice sheet while spreading over the ocean. The friction of these ice shelves with local high spots in the sea floor, or with the sides of embayments, helps restrain the spreading of the ice sheet much as a flying buttress supports a cathedral. The ice shelves are at the melting point where they contact water below, and are relatively low in elevation hence warm above. Ice shelves thus are much more easily affected by climatic warming than are the thick, cold central regions of ice sheets. Rapid melting or collapse of several ice shelves has occurred recently, in response to both atmospheric warming and to intrusion of warmer ocean water, allowing the "gothic cathedrals" behind to spread faster, contributing to sea-level rise. Many additional ice shelves remain that have not changed notably, and these contribute to buttressing of much more ice than was supported by those ice shelves that experienced the large recent changes, so the potential for similar changes contributing to sea-level rise in the future is large.

There are no large ice shelves fed by glaciers in warmer areas, including Alaska, southern Greenland, and elsewhere. Warming of air or water beyond some threshold leads to ice-shelf loss, leaving a calving cliff of the sort that tourists visit on Alaskan cruises or that produce Greenland icebergs that have been filmed and viewed tens of millions of times on video- sharing sites.

Many tourists on Alaskan cruises visit Glacier Bay National Park and Preserve, Alaska. When George Vancouver visited the area in 1794, the bay, which is now about 65 miles long, was completely filled with ice. Less than a century later, when John Muir studied there, most of the bay had lost its ice, as icebergs broke off the front of the glacier at a rate of up to 7 miles per year (Meier and Post, 1987; Post, 1975), and the ice thinned by as much as 1 mile (Larsen et al., 2005). As described by Meier and Post (1987) and many other researchers, ice that ends in the ocean often stabilizes on a local high point or a narrowing of a fjord. It then may remain there for some time despite small climate changes. Under sufficiently large forcing, though, it may retreat rapidly through the deeper or wider part of its valley to the next point of stability, losing ice at a rate that tends to increase with water depth and valley width, but that also depends on other controls. Such behavior has been observed from other Alaskan glaciers, in Greenland, and along the Antarctic Peninsula and elsewhere, and geological evidence indicates that such processes contributed to loss of older ice masses during past, natural warmings.

For Glacier Bay, and many others, the glacier flowed in a relatively narrow fjord. When the ice retreated rapidly from one point of stability to the next, the local changes were spectacular, but the global significance was relatively small. Even in Greenland, the beds of the deep fjords rise inland to near or above sea level, and rapid iceberg calving in deep water cannot discharge most of the ice sheet. Some parts of the Antarctic ice sheet, however, have glaciers that drain large basins rather than narrow fjords, but with ice that is much too thick to float and thus that can raise sea level. Attention is especially, but not uniquely, focused on Thwaites

Glacier in West Antarctica, where sufficient retreat could trigger loss of the marine ice of West Antarctica that would add about 11 feet to the total sea-level rise (Scambos et al., 2017). One attempt to put this process into a model and simulate the future found that, once rapid retreat was initiated, most of West Antarctica's marine ice would be lost over the next century or so (DeConto and Pollard, 2016). That model, though, did not allow icebergs to calve faster than a rate that was based on previous observations in Greenland; Thwaites could retreat to regions where the water is much deeper and the valley much wider than in Greenland, and the tendency for deeper water and wider valleys to have faster calving then allows the possibility of even faster sea-level rise.

The uncertainties about this topic are very large. Thwaites Glacier, and other Antarctic outlets, may resist retreat, or they may retreat with intact ice shelves that limit iceberg calving and thus slow the resulting sea-level rise, and additional stabilizing influences may resist collapse. But, Thwaites may retreat rapidly, and East Antarctic ice may join, with even greater potential to cause rise than has been modeled by, e.g., DeConto and Pollard (2016). The basic physical processes involved have been known for a long time, but properly modeling them is at a very early stage, subject to deep uncertainty.

Uncertainty and Commuting

I am fortunate to ride my bicycle or jog to work; my wife and I share one car, which she mostly drives. But, I have on occasion driven to Washington, DC, and observed automobile commuting. In my experience, a typical commuter on the Beltway heading downtown at rush hour expects notable delays—say, half an hour. The best commute has no delays. Some commuters may waste an hour in traffic. But, occasionally, a commuter is run over by a drunk driver, and ends up in the hospital or worse. When a commuter sets off on a trip, the most-likely future (half an hour in traffic) is close to the "good" end of the possible futures (no traffic, versus run over by drunk driver).

The uncertainties about sea-level rise have a vaguely similar distribution. With warming, ocean water expands, mountain glaciers and the edges of Greenland's ice sheet melt more, and the ocean rises in response. The relevant science, as assessed by the IPCC or other authoritative bodies, gives very high confidence that some additional sea-level rise is already committed as the ice and ocean "catch up" with the warming already caused. Furthermore, additional warming will cause more sea-level rise. The uncertainties include a little less or a little more sea-level rise. But, those projections do not include major instability of Antarctic ice. If "collapse" is triggered in West or East Antarctica, a natural "drunk driver", then the sea-level rise could be much larger and faster than in the most-likely projections, with no real agreement on what might be the worst-case scenario.

Dedicated police officers, watchful bartenders, and other community members help reduce drunk driving, while automotive engineers, highway designers, and additional experts help improve safety in the event of an accident. Society thus has considerable knowledge of ways to reduce risks to commuters. But, we still cannot predict the where and when of every drunk driver. Please note that my next statement is potentially self-serving, because colleagues, students, and I personally enjoy conducting research and may receive funding to do research; but, further research can greatly reduce uncertainties on ice sheets and sea-level rise, providing guidance to policy-makers. Nonetheless, even with well-supported, vigorous research, including ongoing efforts by exceptional colleagues, there may remain some irreducible uncertainty.

Synopsis

With high scientific confidence, human actions are raising the greenhouse-gas concentration of the atmosphere especially by releasing carbon dioxide from fossil-fuel burning, this is having a warming influence on the climate, and the resulting rise in temperature is contributing to sea-level rise by expanding ocean water, melting ice, and changing ice flow.

Some additional sea-level rise is already committed, but human decisions will become increasingly important in determining how much the sea rises over the coming decades and centuries. If Antarctic ice avoids rapid collapse, then uncertainties in projections of sea-level rise are important but not huge, are being addressed by ongoing research, and can be reduced further by planned research. Rapid iceberg calving under too much warming is a well-known but poorly modeled process; if this becomes active in large, deep Antarctic basins, then sea-level rise could be much larger than generally projected, with much greater uncertainties.

References Cited

Alley, R. B., Anandakrishnan, S., Christianson, K., Horgan, H. J., Muto, A., Parizek, B. R., Pollard, D. & Walker, R. T. (2015). Oceanic forcing of ice-sheet retreat: West Antarctica and more. *Annual Reviews of Earth and Planetary Sciences, 43*, 207–231.

American Institute of Physics. (2008). *The Discovery of Global Warming*, by Spencer Weart, http://www.aip.org/history/climate.

Bamber, J. L., Oppenheinmer, M., Kopp, R. E., Aspinall, W. P. & Cooke, R. M. (2019). Ice sheet contributions to future sea-level rise from structured expert judgement. *Proceedings of the National Academy of Sciences of the United States of America, 116*(23), 11195-11200.

CCSP. (2008). *Abrupt Climate Change*. A report by the U.S. Climate Change Science Program and the Subcommittee on Global Change Research [Clark, P.U., A.J. Weaver (coordinating lead authors), E. Brook, E.R. Cook, T.L. Delworth, and K. Steffen (chapter lead authors)]. U.S. Geological Survey, Reston, VA, 244 pp.

CCSP. (2009). *Past Climate Variability and Change in the Arctic and at High Latitude.* A report by the U.S. Climate Change Program and Subcommittee on Global Change Research [Alley, R.B., Brigham-Grette, J., Miller, G.H., Polyak, L., and White, J.W.C. (coordinating lead authors). U.S. Geological Survey, Reston, VA, 461 pp.

Church, J. A., Gregory, J. M., Huybrechts, P., Kuhn, M., Lambeck, C., Nhuan, M. T., Qin, D. & Woodworth, P. L. (2001). Changes in sea level. in: J.T Houghton, Y. Ding, D.J.Griggs, M. Noguer, P.J. Van der Linden, X. Dai, K. Maskell, and C.A. Johnson(eds.): Climate Change 2001: The Scientific Basis: Contribution of Working Group Ito the Third Assessment Report of the Intergovernmental Panel on Climate Change, Cambridge University Press (Cambridge, New York), 639-694.

Church, J. A., Clark, P. U., Cazenave, A., Gregory, J. M., Jevrejeva, S., Levermann, A., Merrifield, M. A., Milne, G. A., Nerem, R. S., Nunn, P. D., Payne, A. J., Pfeffer, W. T., Stammer, D. & Unnikrishnan, A. S. (2013). Sea Level Change. In: *Climate Change 2013: The Physical Science Basis. Contribution of Working Group I to the Fifth Assessment Report of the Intergovernmental Panel on Climate Change* [Stocker, T.F., D. Qin, G.-K. Plattner, M. Tignor, S.K. Allen, J. Boschung, A. Nauels, Y. Xia, V. Bex and P.M. Midgley (eds.)]. Cambridge University Press, Cambridge, United Kingdom and New York, NY, USA.

Cuffey, K. M. & Paterson, W. S. B. (2010). *The Physics of Glaciers, Fourth edition*. Oxford. DeConto, R.M and D. Pollard. 2016. Contribution of Antarctica to past and future sea-level rise. *Nature, 531*(7596), 591–597. doi:10.1038/nature17145.

Garner, A. J., Weiss, J. L., Parris, A., Kopp, R. E., Horton, R. M., Overpeck, J. T. & Horton, B. P. (2018). Evolution of 21st century sea level rise projections. *Earth's Future*, 6, 1603–1615.

Griggs, J. A. & Harries, J. E. (2007). Comparison of spectrally resolved outgoing longwave radiation over the tropical Pacific between 1970 and 2003 using IRIS, IMG, and AIRS. *Journal of Climate*, *20*, 3982–4001.

Harries, J. E., Brindley, H. E., Sagoo, P. J. & Bantges, R. J. (2001). Increases in greenhouse forcing inferred from the outgoing longwave radiation spectra of the Earth in 1970 and 1997. *Nature*, *410*, 355–57.

The IMBIE Team. (2018). Mass balance of the Antarctic ice sheet from 1992 to 2017. *Nature*, *558*, 219-222. doi: 10.1038/s41586-018-0179-y.

Larsen, C. F., Motyka, R. J., Freymuller, J. T., Echelmeyer, K. A. & Ivins, E. R. (2005). Rapid viscoelastic uplift in southeast Alaska caused by post-Little Ice Age glacial retreat. *Earth and Planetary Science Letters*, *237*(3-4), 548-560, doi: 10.1016/j.epsl.2005.06.032.

Lemke, P., Ren, J., Alley, R. B., Allison, I., Carrasco, J., Flato, G., Fujii, Y., Kaser, G., Mote, P., Thomas, R. H. & Zhang, T. (2007). Observations: Changes in Snow, Ice and Frozen Ground, in Solomon, S., D. Qin, M. Manning, Z. Chen, M. Marquis, K. B. Averyt, M. Tignor and H. L. Miller [eds.], *Climate Change 2007: The Physical Science Basis. Contribution of Working Group I to the Fourth Assessment Report of the Intergovernmental Panel on Climate Change* [Cambridge University Press, New York.

Levermann, A., Clark, P. U., Marzeion, B., Milne, G. A., Pollard, D., Radic, V. & Robinson, A. (2013). The multimillennial sea-level commitment of global warming. *Proceedings of the National Academy of Sciences of the United States of America*, *110*(34), 13745-13750, https://doi.org/10.1073/pnas.1219414110.

Masson-Delmotte, V., Schulz, M., Abe-Ouchi, A., Beer, J., Ganopolski, A., González Rouco, J. F., Jansen, E., Lambeck, K., Luterbacher, J., Naish, T., Osborn, T., Otto-Bliesner, B., Quinn, T., Ramesh, R., Rojas, M., Shao, X. & Timmermann, A. (2013). Information from Paleoclimate Archives. In: *Climate Change 2013: The Physical Science Basis. Contribution of Working Group I to the Fifth Assessment Report of the Intergovernmental Panel on Climate Change* [Stocker, T.F., D. Qin, G.-K. Plattner, M. Tignor, S.K. Allen, J. Boschung, A. Nauels, Y. Xia, V. Bex and P.M. Midgley (eds.)]. Cambridge University Press, Cambridge, United Kingdom and New York, NY, USA.

McGranahan, G., Balk, D. & Anderson, B. (2007). The rising tide: Assessing the risks of climate change and human settlements in low elevation coastal zones. *Environment and Urbanization*, *19*, 17–37.

Meier, M. F. & Post, A. (1987). Fast tidewater glaciers. *Journal of Geophysical Research*, *92*(B9), 9051-9058.

National Research Council, National Academy of Sciences, National Academies Press, Washington, DC:

1975: United States Committee for the Global Atmospheric Research Program, *Understanding Climate Change: A Program for Action*

1979: Ad Hoc Study Group on Carbon Dioxide and Climate, *Carbon Dioxide and Climate: A Scientific Assessment*

2001: Committee on the Science of Climate Change, *Climate Change Science: An Analysis of Some Key Questions*

2002: Committee on Abrupt Climate Change, *Abrupt Climate Change: Inevitable Surprises*

2006: Committee on Surface Temperature Reconstructions for the Last 2,000 Years, *Surface Temperature Reconstructions for the Last 2,000 Years*

2008: *Understanding and Responding to Climate Change: Highlights of National Academies Reports*

2010a: Committee on Stabilization Targets for Atmospheric Greenhouse Gas Concentrations, *Climate Stabilization Targets: Emissions, Concentrations, and Impacts over Decades to Millennia*

2010b: Panel on Advancing the Science of Climate Change, *Advancing the Science of Climate Change*

2013: Committee on Understanding and Monitoring Abrupt Climate Change and Its Impacts, *Abrupt Impacts of Climate Change: Anticipating Surprises*

Nerem, R. S., Beckley, B. D., Fasullo, J. T., Hamlington, B. D., Masters, D. & Mitchum, G. T. (2018). Climate-change-driven accelerated sea-level rise detected in the altimeter era. *Proceedings of the National Academy of Sciences of the United States of America*, *115*(9), 2022-2025.

Post, A. (1975). *Preliminary hydrology and historic terminal changes of Columbia Glacier, Alaska.* U.S. Geological Survey Hydrological Investigations Atlas.

Rahmstorf, S., Cazenave, A., Church, J. A., Hansen, J. E., Keeling, R. F., Parker, D. E. & Somerville, R. C. J. (2007). Recent climate observations compared to projections. *Science*, *316*, p. 709.

Scambos, T., Bell, R. E., Alley, R. B., Anandakrishnan, S., Bromwich, D. H., Brunt, K., Christianson, K., Creyts, T., Das, S., DeConto, R., Dutrieux, P., Fricker, H. A., Holland, D., MacGregor, J., Medley, B., Nicholas, J. P., Pollard, D., Siegfried, M. R., Smith, A. M., Steig, E. J., Trusel, L. D., Vaughan, D. G. & Yager, P. L. (2017). How Much, How Fast?: A Review and Science Plan for Research on the Instability of Antarctica's Thwaites Glacier in the 21st century. *Global and Planetary Change*, *153*, 16-34.

Slater, T. & Shepherd, A. (2018). Antarctic ice losses tracking high. *Nature Climate Change*, *8*, 1024-1026.

White, J. W. C., Alley, R. B., Brigham-Grette, J., Fitzpatrick, J. J., Jennings, A. E., Johnsen, S. J., Miller, G. H., Nerem, R. S. & Polyak, L. (2010). Past rates of climate change in the Arctic. *Quaternary Science Reviews*, *29*(15-16), 1716-1727.

Short Biography

Dr. Richard Alley (PhD 1987 University of Wisconsin; MSc 1983 and 1980 The Ohio State University, all in Geology) is Evan Pugh University Professor of Geosciences and Associate of the Earth and Environmental Systems Institute at the Pennsylvania State University. He has spent more than 40 years studying the great ice sheets to help predict future changes in climate and sea level, and has made four trips to Antarctica, nine to Greenland, and additional expeditions to Alaska and elsewhere. He has been honored for research (including election to the US National Academy of Sciences and Foreign Membership in the Royal Society), teaching (including Penn State's highest teaching award), and service. Dr. Alley participated in the UN Intergovernmental Panel on Climate Change (co-recipient of the 2007 Nobel Peace Prize), and has provided requested advice to numerous government officials in multiple administrations including a US Vice President, Presidental Science Advisors, and committees and individual members of the US Senate and House of Representatives. He has authored or coauthored more than 300 refereed scientific papers. He was presenter for the PBS TV miniseries on climate and energy *Earth: The Operators' Manual*, and author of the book. His popular account of climate change and ice cores, *The Two-Mile Time Machine*, was Phi Beta Kappa's science book of the year. Dr. Alley is happily married with two grown daughters, two stay-at-home cats, a bicycle, and a pair of soccer cleats.

Dr. Richard B. Alley
Evan Pugh University Professor
Department of Geosciences, and Earth and Environmental Systems Institute
The Pennsylvania State University

Chairwoman Johnson. Thank you very much. Dr. Bell.

TESTIMONY OF DR. ROBIN E. BELL, LAMONT RESEARCH PROFESSOR, LAMONT-DOHERTY EARTH OBSERVATORY, COLUMBIA UNIVERSITY

Dr. BELL. Thank you very much. Chairwoman Johnson, Ranking Member Lucas, Members of the Committee, I'm very pleased to be here today. I'm going to take you on a visual tour because the ice sheets are beautiful. I think that's why we all study them. And we want to share a little bit of that beauty with you, so this is a picture of what Antarctica looks like. And just to give you a sense of scale, this is a huge iceberg with tiny scientists in front of it.

What I'm going to show you today is the evidence for change. I will tell anyone this who stops me anywhere on the street—you stop me on my electric motorcycle, I'm going to tell you this story. There are three signs of change—three really clear signs that the ice sheets are speeding up and changing. One is they're moving faster. In the 1990s, they were moving 1 mile a year. In the 2000s, they're moving 2 miles a year. They've doubled in speed.

Ice happens to be like the mozzarella cheese on top of your pizza, so when you bite into the cheese and stretch it, it gets thinner. So the second measurements we've made is by zapping the ice sheet with a laser, and that's what you see forming is that yellow on the surface is actually where the elevation, just like the cheese is getting stretched, the ice sheet is getting stretched, that's more than half a football field of stretching where the ice sheet is getting lower, second measurement.

Our third measurement is one we make and NASA makes with partners—makes from space. Can we turn the video on—animation on, please? You will see that this is Antarctica again—now we're looking at a whole map of Antarctica, and you're going to see a red dot develop. And what that red dot is showing we're actually losing mass. And remember I showed you it sped up, it lowered. This is a different measurement. This is basically the ice sheet on the bathroom scale. And what you can see is the ice sheet is losing mass predominantly in that place that Richard referred to, the place that's furthest north and exposed to the warming ocean. The ice sheet is losing mass.

We could show you the same things for Greenland, three very clear signals, kind of the scientific gold standard. We like to make independent measurements. This is the evidence that the ice sheets are changing.

What does it mean? We go next to NOAA (National Oceanic and Atmospheric Administration) and we look at NOAA's global collection of tide gauges—so these are really high-tech instruments. They're like pipes stuck in the water, OK? But they measure the tides going up and down and up and down, and they measure storms—the tide levels go way up, 12 feet in New York during Sandy. But you can see most of those are going up. Sea level almost everywhere on the planet is going up except where the planet is still recovering from the ice sheet that was more than 20,000 years ago and it's bouncing back up like a mattress. But this predominant signal globally is it's going up.

There's even one of those fancy pipes right here in Southeast, Washington, and that record goes back to about when my dad was born.

And so since my dad was born right here in Washington, D.C., sea level has gone up a foot. And we're using Beth for scale here. Beth is, for today's purposes, 2 meters or 6 feet roughly. And you can see, sea level has risen almost a foot, almost to my knee— I like to think of it—I put my hand on my leg because then I realize what it really means. That's how far sea-level has come up since my dad was born.

So what does that mean? We are working on this problem—I'm back to the uncertainty question. Can I tell you how much sea-level is going to go up in the next hundred years? We are working on it as hard as we can. This is just a range of forecasts published this year. You can see the results—they are spread. This is again Beth for scale, about 6 feet. They range a lot. But when we looked— that's what we're working on is how to be able to tell our communities how much is sea level to go up in the next hundred years because that's what we're building infrastructure. The big bridge we just spent $4 billion on close to my house needs to know what we're going to plan for sea level. Are we going to plan for a couple feet or a lot more?

So when we look at the glacier melt budget altogether, Antarctica is in the next hundred years is on the order of maybe over our knees, maybe a little bit more. Greenland is going to be in there, too. We're going to have warming oceans, and we're going to have mountain glaciers. And while I have this as roughly 4 feet, 3 feet, we don't know. This is cutting-edge research.

And what can we do to improve it? There's a priority of three ideas in my mind—there's three important things to do. One is get up close and personal to the ice sheets. We need to understand better how the ice sheets work so we can improve our models. We used to not be able to have very good models of weather. We do much better now. So number one is get up close and personal.

Second is we need to invest in the workforce. Right now, there are 1,400 scientists at the AGU who are affiliated with ice. Do you know there are 140,000 people enrolled in law school every year? We just don't have enough people working on this. We need more scientists, engineers, educators, creative minds like Maria over there. We need to talk her into studying ice somehow.

And we also need to look at how convergent science works. We need to figure out how to pull together the work that we do, which is on the polar caps, to what's happening at the coastlines around the planet because we kind of need an ice sheet person in every community because we need to understand what the community needs to respond to.

So am I hopeful? Yes, I am hopeful because we are in a unique place as a species that we know how the ice sheet works, we know how sea level rises—we are understanding how our planet works. And we, as scientists, we're all members of the American Geophysical Union. We're actually putting our money where our mouth is. We have a building here in Washington that we just renovated, so it is the first net-zero renovation building in Washington. That means we're taking less energy than we are generating, more energy than we are using to run this building. We'd love to have you come visit. And we're also very happy to look forward and that this is a time for action among all of us, and we need to bring everybody to the table. Thank you very much.

[The prepared statement of Dr. Bell follows:]

Written Testimony of Professor Robin Elizabeth Bell, PGI Lamont Research Professor, Lamont-Doherty Earth Observatory of Columbia University, President American Geophysical Union.

Earth's Thermometers: Glacial and Ice Sheet Melt in a Changing Climate for Committee on Science Space and Technology, U.S. House of Representatives, July 11, 2019.

Introduction

Chairwoman Johnson, Ranking Member Lucas, Members of the House Science Committee, thank you for inviting me and my esteemed colleagues to speak today on this important topic- the melting ice on our home planet. This change is happening at the ends of our planet but is lapping at our doorsteps now. I am Robin Elizabeth Bell, and I am the PGI Lamont

Research Professor at Lamont-Doherty Earth Observatory of Columbia University and a member of the Earth Institute faculty. At Lamont, I direct programs in ice sheet dynamics, lead efforts to develop innovative technology and work to improve the scientific culture especially for women. I have led ten major expeditions to the polar regions, to both Greenland and Antarctica resulting in discoveries ranging from active volcanism beneath the West Antarctic Ice sheet, to large deep lakes encased by two miles of ice to hidden mountain ranges buried by ice where water under the pressure of thick ice is forced uphill. I was the first women to chair the National Academy of Science's Polar Research Board (2002-2008) where I was instrumental in launching the International Polar Year 2007-9 that brought together over 50,000 scientists from around the globe. The International Polar Year fostered major expeditions, new international collaborations and discoveries that were only possible because of the partnerships between 60 nations. The polar regions are still a challenging place to work and science remains an international team sport. I also co-chaired the recent National Academy report *A Strategic Vision for NSF Investments in Antarctic and Southern Ocean Research* (2015) that set the priorities for Antarctic science. This report identified changing ice as the highest priority science in Antarctica. Recently, I chaired the National Academy of Science Review of the Draft Fourth National Climate Assessment – a comprehensive undertaking by U.S. scientists and citizens documenting the impacts of climate change around the country (USGCRP 2018). Currently, I have the great honor of serving as the President of the American Geophysical Union, or AGU as we all refer to it. Formed 100 years ago, AGU is the society of over 60,000 Earth and space scientists from around the globe who together promote discovery for the benefit of humanity. Today, I am honored to speak about the changing ice sheets with you. Thirty years ago when I first flew over Antarctica in a Naval Research Laboratory P-3 it seemed unimaginable to me that the vast ice sheet below could change. Now we know those white expanses are changing and these changes matter to our homes and communities around the globe. The changing polar ice is tightly linked to the changing coastlines. Although I speak to you today in my capacity as a private citizen, my testimony is based on my decades of experience studying our planet's ice.

Evidence for Changing Ice

The surprising wakeup call for the polar science community came in early 2002. This buzzing alarm came from the Antarctic Peninsula, the part of Antarctica that is the furthest north, jutting towards South America. This is also the destination of Antarctic cruises which over 14,000 Americans visit each year The Antarctic Peninsula is where global temperatures have risen the most - more than 7°F over 50 years. We had thought ice sheets and the state-sized ¼-1/2 mile thick pieces of floating ice that pin them in place changed really slowly. These floating extensions of continental ice sheets are called ice shelves. But by 2002, warming temperatures had started to produce more meltwater on top of the ice. The floating Larsen B Ice Shelf, the size of Rhode Island, developed hundreds of lakes. Suddenly the ice shelf disintegrated into thousands of icebergs over the course of two weeks (Scambos, Hulbe et al. 2003). The change occurred before our very eyes. The Larsen B ice shelf had been in place for over 10,000 years (Domack, Duran et al. 2005). Once the floating ice shelf disintegrated, the glaciers that flowed into the ice shelf sped up, pushing more ice into the ocean (Rignot, Casassa et al. 2004, Scambos, Bohlander et al. 2004). Glaciers are the earth's conveyor belts delivering ice to the ocean and an ice shelf controls the speed– if an ice shelf collapses, the conveyor belt speeds up. The satellite images of this collapse were printed in major newspapers around the globe (for example: The Charleston Gazette, St Louis Post, The Gazette – Ft. Wayne Indiana, The Patriot – Harrisburg, PA, Chicago Tribune, Rocky Mountain News, The Economist, The Wall Street Journal, The New York Times, Toronto Star, Calgary Herald, The Press - New Zealand, Belfast Telegraph, The Australian, China Daily, The Statesman- India). Suddenly, changing ice was newsworthy. Together scientists and the public from Harrisburg to India learned Antarctic ice could change faster than we imagined. The Antarctic conveyor belt had sped up. For the first time many around the world saw the link between blue meltwater on the ice shelf surface, the glacier conveyor belt speeding up and sea level rising.

Over the ensuing decades, the evidence for the changing ice on our planet has become very clear. I will focus on the grounded ice, the large ice sheets in Antarctica and Greenland where thick ice, in places over two miles thick, rests on solid ground although the ground may be well below sea level. Melting these ice sheets will raise sea level around the globe. Antarctica holds 200 feet of potential sea level rise and Greenland 20 feet of sea level rise– although no scientists are suggesting they will completely disappear any time soon. These very thick ice sheets are distinct from the relatively thin floating sea ice (around 10 feet) that covers much of the Arctic Ocean and rings the Southern Ocean close to Antarctica. Sea ice is like the layer of ice cubes floating in a punch bowl. The Arctic sea ice has been steadily shrinking over the past two decades and recently the Antarctic sea ice has begun to retreat. Changing sea ice shifts the Earth's albedo and weather patterns, impacts food available to wildlife from penguins to polar bears, and opens new shipping routes. But shrinking sea ice itself will not cause sea level to rise, since sea ice is already floating in the water. The major source of future sea level rise are the grounded ice sheets. Melting ice sheets are like the kid with a new twenty-pound bag of ice at the picnic who pours the entire bag into the bowl without thinking. The glaciers are conveyor belts of ice being delivered to the ocean, and we see them speeding up.

I often get asked "do you believe the ice is changing?" My response is – changing ice is not a belief but knowledge that emerges from three independent observations. These independent observations are primarily based on satellite measurements enabled by NASA working with other space agencies around the globe. The first measurement is how fast the ice moves. Several parts of the Antarctic Ice Sheet (key parts of the conveyor belts) have doubled their speed in the past two decades, showing that the ice is speeding up (Rignot, Mouginot et al. 2011). The second measurement is the height of the ice surface, and is made using laser and radar instruments from a satellite or aircraft. In the same places where the ice is speeding up the ice surface is getting lower. Ice, like mozzarella cheese atop a pizza, is getting thinner and lower because it is stretching. The third measurement is ice sheet mass, or weight, which is calculated from observations from a pair of identical satellites chasing each other and measuring changes in the gravity field (Velicogna, Sutterley et al. 2014, Harig and Simons 2015). In the same places that the ice is speeding up and lowering, it is losing mass. These three measurements together demonstrate in more detail than ever before how the ice in Greenland and Antarctica is changing.

Scientists from around the globe have used these three key observations to quantify how fast the ice sheets are changing. To quantify the change over a large continent like Antarctica, the size of the lower 48 states, requires careful examination of each measurement and resolving issues such as how the snow that falls on Antarctica turns into ice. After much lively debate and testing of assumptions by a team of 77 scientists from around the world, the clear signal is that Antarctica is losing ice, as is Greenland. The current mass loss from the ice sheets is contributing one millimeter of sea level rise globally each year (Shepherd, Ivins et al. 2018) although this rise is not evenly distributed around the globe. Antarctica is now losing mass at twice the rate it was in the 1990s. For these calculations, the team broke Antarctica up into three parts, the Peninsula where the Larsen B Ice Shelf was; West Antarctica, the ice sheet that rests on low-lying topography and is exposed to changes in the ocean temperature and East Antarctica, the large ice sheet where the South Pole is that sits on higher topography. Each region stores different amounts of ice, has a different history and a different susceptibility to a warming world. The West Antarctic Ice Sheet is the most susceptible to warming oceans and atmosphere as it sits lower and is in direct contact with the ocean. West Antarctica is where the greatest changes have occurred over the past decade. Most of the 0.3 inches (8 mm) of sea level rise from Antarctica in the last decade has come from West Antarctica. This region was the highest priority in the 2015 National Academy report *A Strategic Vision for NSF Investments in Antarctic and Southern Ocean (National Academies of Sciences and Medicine 2015).*

Evidence for Changing Coastlines

Why did the changing ice emerge as the highest priority in the National Academy Report? We are beginning to see the melting ice, including from Antarctica, at the tide gauges along our coastline. Globally, average sea level has risen 8-9 inches since 1880, with the global rise since 1993 being 3 inches (Hay, Morrow et al. 2015, Nerem, Beckley et al. 2018). Right here at the dock along the bike path in Southeast Washington sea level has risen a foot in since 1919 (https://tidesandcurrents.noaa.gov/sltrends). I put my hand on my leg just below my knee and realize the water level has risen that far since my father was born.

At most locations around the globe sea level is rising now, although the ocean turns out to be more complicated than the punch in the punchbowl. At a few locations, sea level is actually falling. Three major components make up the change at an individual coastal city: the change in ocean temperature, the melting ice and whether the land the city rests on is rising or sinking. Up to now the warming of the ocean waters by 1.3°F since 1960 is the major signal that has appeared at our coasts. But, melting ice has the greatest potential for new rapid sea level rise globally. To complicate things further, the melting ice contribution to changes sea level is modulated by the self-gravitation of the ice sheets. Already the modulation of the impact of melting ice in Greenland by the self-gravitation is apparent in the tide gauges along the east coast of the United States. Because of this gravitational effect, sea level is rising faster in the Southeastern US than in New England. Atop these signals are local impacts. The land cities and towns are built on can be rising or sinking, impacting local sea level. In cities like Juneau and Stockholm (Milne, Davis et al. 2001) the land is rising due to the loss of ice 20,000 years ago while cities like Norfolk, Virginia and New Orleans are sinking due to removal of groundwater (Sweet, Kopp et al. 2017). Every community is going to see a different future sea level depending the ocean temperature, the changing ice, and whether the land is rising or falling. Linking the changing ice to the changing coastlines is a challenge that will require collaboration from the ice to the shorelines.

Impacts of Changing Coastlines Now and Looking Ahead

So we have begun to witness the melting ice and see the impact along our shorelines. The higher sea level made the impact of recent major storms like Maria, Harvey, Irma and Sandy more devastating. For example, close to my home 30 miles from the Atlantic Ocean they used bulldozers to clear boats from the roads after Superstorm Sandy. Because of the sea level rise over the past century, 45,000 more people were impacted by Sandy's flooding. The impact of rising sea level is not just during major storms. All around the US we are seeing increased nuisance flooding. Nuisance flooding is called sunny day flooding where high tides in fair weather make it difficult to get home because the roads are flooded. Miami and Norfolk are both experiencing this and are working to adapt to this. Scientists are working to provide these cities with the forecasts of future sea level they need to adapt.

Looking Ahead: Current Ice Sheet Change Projections

Looking ahead the scientific community is scrambling to provide answers to how fast and how much will sea level rise in each community from ice sheet melt. Suddenly city managers, architects, reinsurance companies and resiliency officers care about Antarctic ice. The efforts to answer the how-fast-how-much question range from simple exercises to frame the problem to quizzing experts locked in a room (Bamber and Aspinall 2013) to probabilistic projections (Edwards, Brandon et al. 2019) and full-blown ice sheet models (Feldmann and Levermann 2015, DeConto and Pollard 2016). These models are like weather models only for ice. In contrast to weather and hurricane models, these models are still in the early stages of development. Ice sheet modeling scientists have made big advances in these efforts, such as figuring out how to capture mathematically the changing forces when ice goes afloat and using the latest supercomputing resources to allow the models to include many of the important stresses at play within the ice. The ice sheet models are now linked to different futures, whether temperatures go up a little, a lot or a huge amount. These different futures will be determined by how much CO2 we release into the atmosphere. The science community is working through this collaboratively and through peer review, the way good science happens. An idea is published, the community tests it and new ideas are advanced. Since scientists have never watched an ice sheet disappear, we use records from the past. We know sea level rises when temperature rise --- Miami is built on rocks formed in a shallow sea very similar to the Bahamas today. The hills of Miami formed 120,000 years ago when the planet was warm and sea level was 19-30 feet higher than it is now. The other point we use to calibrate our models is from three million years ago, the last time CO_2 was as high as it is now sea level was 19-65 feet higher than it is now.

The challenges the scientists working on the models face include that is that we are still learning so much about how ice sheets work. For example, while we are all familiar with how water flows across our familiar landscape, we are just now working to understand what happens when water collects on Antarctica. Greenland wears a necklace of blue ponds every summer and has water hidden in crevasses and in the snow. What happens if Antarctica warms until it looks like Greenland (Bell, Banwell et al. 2018)? Will all the new water make the remaining ice shelves disintegrate like the Larsen B, triggering more glaciers/conveyor belts to accelerate, or will rivers form atop the ice (Kingslake, Ely et al. 2017)? Will we will see giant ice cliffs that become unstable causing a sudden runaway collapse of the ice switching the glacier conveyor belts to hyper-fast? These are the ice processes that might produce drastically accelerate sea level rise. Models with lots of meltwater and collapsing cliffs predict close to six feet of sea level rise from Antarctica by 2100. More recent publications suggest that the number might be closer to 1-1.5 feet (45 cm). As we discover new important processes and discover more, these numbers will change. Our knowledge-base and our models are evolving. My family has a boat on the Hudson and we worry about hurricanes every summer. Thirty years ago the hurricane models could not tell us whether the hurricane was going to hit Maine or our New York home, now we can plan much better. We knew Sandy was possibly coming ten days out and were able to prepare. The improvement in hurricane prediction illustrates that the ice predictions can improve if we work on it by building our knowledge base, deepening the bench of scientists and fostering interdisciplinary and international collaborations.

Three Essentials to Improve Ice Sheet Melt Projections

The Antarctic melt projections for 2100 range from just below my knee or over my head, or, quantitatively 1-6 feet. How can we narrow down this answer about how Antarctica will melt in the coming decades? My neighbors are asking me. There are three critical things essential to improving the predictions: knowledge of processes (or how ice sheets work), people (to explore, discover, model and communicate, and fostering collaboration: 1) Processes: We have never witnessed an ice sheet collapse and improving our predictions requires getting up close and personal with the ice sheets to better understand how ice sheets work and intense efforts to decide how best to describe these processes in ice-sheet models. 2) People: The community studying ice around the world has grown but the community is still really small. 3) Collaboration: Because changing ice is controlled by the ocean, the atmosphere, the underlying geology and ice physics and Antarctica are huge, this work requires collaboration across disciplines and nations.

Our understanding of the process of how ice sheets work has made huge advances. Prior to the International Geophysical Year in 1958, we did not even know how much ice there was in Antarctica. By the 1980s we began to understand why those giant conveyor belts of ice can deliver so much ice to the ocean (Alley 1986). These conveyor belts can be over 60 miles wide and in Antarctica move up to about 1.5 miles per year. In Greenland the conveyor belts move even faster – more than 7.5 miles a year. In the 1990s we began both to drill through the ice sheet and to study extensive regions with aircraft and we discovered that the geology underneath matters. In the 2000s we realized there were extensive networks of water beneath the ice including large lakes, one the size of New Jersey (Kapitsa, Ridley et al. 1996), smaller lakes that will slowly fill and drain (Fricker, Scambos et al. 2007), and water networks that move the water. Where the water goes matters because the water is part of the basal lubrication system. Some of the big unknowns include: what is happening in this hidden environment beneath the ice, how will the warming ocean and atmosphere attack the ice sheet and will surface water trigger collapse of all the major ice shelves?

We have discovered a lot but there remains a lot to be learned. We as a species have lived with changing weather and have a deep knowledge of weather systems behave. Our grandmothers understood the wispy angular clouds they called mare's tails meant rain soon, but we can now predict to the hour when the rain will arrive. We as a species have far less experience with collapsing ice sheets. To improve our models, we must get up close and personal with the ice sheets. The satellite record has clearly shown us that change is happening but it is the work in Antarctica from surface ships and aircraft that is essential to foster the advances in understanding of how ice sheets work that will improve our projections. NASA's Operation Icebridge is an example of the importance of comprehensive imaging of the ice sheets that fostered a new norm of freely available open data. The National Science Foundation has responded to the 2015 National Academy report by launching a major program collaboratively with the United Kingdom's Natural Environment Research Council (NERC), the International Thwaites Glacier Collaboration. Thwaites Glacier, one of the largest conveyor belts, is considered one of the most unstable pieces of ice on the planet. Thwaites Glacier is wide and is perched on a topographic ridge where the warming ocean is known to be thinning the ice. Because this glacier can deliver a lot of ice to the ocean fast and because it is already showing signs of thinning and shrinking it is a major threat and a high priority (NAS). The major NSF/NERC initiative (Scambos, Bell et al. 2017) this as an example of the type of work that is essential to launch around all of Antarctica. Advancing the basic understanding of how the ice sheets work and the processes that control their melting, will improve our predictions. Think of Antarctic scientists as the hurricane hunters for sea level.

The second critical need to improve our projections is people. As President of AGU and as former President of the Cryosphere Section (the best job title ever), I am acutely aware of how small our community is. Now, the AGU Cryosphere section has 1,492 members. This number includes scientists from around the globe studying ice, snow and sea ice. To put that in perspective in 2010, there were about 140,000 people enrolled in law school in the US. In a single year, 100 times more people were studying law than the entire global community studying changing ice. There is an acute personnel problem. We need more scientists working on this problem if we are to improve our projections. Science and the science of melting ice from the Arctic to the South Pole must be an open welcoming community. The science is remarkable and the discoveries to be made remarkable. We have barely started to scratch the surface of the ice sheets.

The third need is to fully embrace ice as part of the changing earth and enable truly convergent work. When your child is in the hospital with a sudden ailment you really want the specialists to be working together to provide the best care. The ice community is coming to the realization that we need to take a similar approach. We recently completed ROSETTA, a study of the largest ice shelf in Antarctica, the Ross, just a little smaller than Texas. Using Recovery Act funding, in partnership with the New York Air National Guard, we repurposed military imaging technology for ice studies. After three years of flying the IcePod over the Ross Ice Shelf we realized that it was impossible to understand how the ice will melt without bringing all the specialists to the table. We learned that the geology is in essence protecting that sector of West Antarctica from the warming global ocean but the vulnerability is to heat pumped under the ice shelf from the shallow ocean waters by strong winds (Tinto, Padman et al. 2019). It took scientists from many disciplines working together on the same data sets to converge on these complex processes. We were acting like that team of specialists working together for the good for a patient. It is essential to foster this convergent work for the planet and our species. To move the Antarctic work forward will require interdisciplinary and international collaboration as fostered by NSF in the ITGC program but on a larger scale. Ice science must also be more tightly linked to our changing coastlines so each community will know how to respond and adapt. I am hopeful. With investment the hurricane forecasts have improved. We can improve the melt forecasts and provide better information to our neighbors.

Thank you for inviting me to speak today. I am heartened that the House Science Committee is considering this very important issue. We are very fortunate as a species to have the capacity to see how our home planet works and have the capacity to address this issue both scientifically and technically. If we continue to foster collaborative science across disciplines, the science community will be able to provide our communities with accurate projections of how sea level will rise. If ice scientists work with coastal scientists, we can develop tailored projections for each community. I am also heartened as I see individuals, communities, state governments and professional societies taking action to reduce the underlying cause of the changing ice – our greenhouse gas emissions. The AGU community is very proud of our headquarters on Florida Avenue. This building, long known for the planets in the sidewalk, is Washington DC's first net-zero emissions building renovation. Reaching net zero required multiple technologies from solar panels to heat exchange with the sewer system to green walls. Similarly, using the same multi-pronged strategy, we as a species can address the issue of climate change and ice melt with broad concerted efforts, from individuals, communities and governments.

References: R.E. Bell House Testimony

Alley, R. B. (1986). "Deformation of till beneath ice stream B, West Antarctica." *Nature*, *322*, 57-59.

Bamber, J. L. & Aspinall, W. (2013). "An expert judgement assessment of future sea level rise from the ice sheets." *Nature Climate Change*, *3*(4), 424.

Bell, R. E., Banwell, A. F., Trusel, L. D. & Kingslake, J. (2018). "Antarctic surface hydrology and impacts on ice- sheet mass balance." *Nature Climate Change*, *8*(12), 1044-1052.

DeConto, R. M. & Pollard, D. (2016). "Contribution of Antarctica to past and future sea-level rise." *Nature*, *531*, 591–597.

Domack, E., Duran, D., Leventer, A., Ishman, S., Doane, S., McCallum, S., Amblas, D., Ring, J., Gilbert, R. & Prentice, M. (2005). "Stability of the Larsen B ice shelf on the Antarctic Peninsula during the Holocene epoch." *Nature*, *436*(7051), 681-685.

Edwards, T. L., Brandon, M. A., Durand, G., Edwards, N. R., Golledge, N. R., Holden, P. B., Nias, I. J., Payne, A. J., Ritz, C. & Wernecke, A. (2019). "Revisiting Antarctic ice loss due to marine ice-cliff instability." *Nature*, *566*(7742), 58.

Feldmann, J. & Levermann, A. (2015). "Collapse of the West Antarctic Ice Sheet after local destabilization of the Amundsen Basin." *Proceedings of the National Academy of Sciences*, *112*(46), 14191-14196.

Fricker, H. A., Scambos, T., Bindschadler, R. & Padman, L. (2007). "An active subglacial water system in West Antarctica mapped from space." *Science*, *315*, 1544-1548.

Harig, C. & Simons, F. J. (2015). "Accelerated West Antarctic ice mass loss continues to outpace East Antarctic gains." *Earth and Planetary Science Letters*, *415*, 134-141.

Hay, C. C., Morrow, E., Kopp, R. E. & Mitrovica, J. X. (2015). "Probabilistic reanalysis of twentieth-century sea- level rise." *Nature*, *517*(7535), 481.

Kapitsa, A., Ridley, J. K., Robin, G. d. Q., Siegert, M. J. & Zotikov, I. (1996). "Large deep freshwater lake beneath the ice of central East Antarctica." *Nature*, *381*, 684-686.

Kingslake, J., Ely, J. C., Das, I. & Bell, R. E. (2017). "Widespread movement of meltwater onto and across Antarctic ice shelves." *Nature*, *544*(7650), 349.

Milne, G. A., Davis, J. L., Mitrovica, J. X., Scherneck, H. G., Johansson, J. M., Vermeer, M. & Koivula, H. (2001). "Space-geodetic constraints on glacial isostatic adjustment in Fennoscandia." *Science*, *291*(5512), 2381-2385.

National Academies of Sciences E. and Medicine. (2015). *A Strategic Vision for NSF Investments in Antarctic and Southern Ocean Research.* Washington, DC, The National Academies Press.

Nerem, R., Beckley, B., Fasullo, J., Hamlington, B., Masters, D. & Mitchum, G. (2018). "Climate-change–driven accelerated sea-level rise detected in the altimeter era." *Proceedings of the National Academy of Sciences*, 201717312.

Rignot, E., Casassa, G., Gogineni, P., Krabill, W., Rivera, A. & Thomas, R. (2004). "Accelerated ice discharge from the Antarctic Peninsula following the collapse of Larsen B ice shelf." *Geophysical Research Letters*, *31*(18).

Rignot, E., Mouginot, J. & Scheuchl, B. (2011). "Ice flow of the Antarctic ice sheet." *Science*, *333*(6048), 1427-1430.

Scambos, T., Hulbe, C. & Fahnestock, M. (2003). "Climate-induced ice shelf disintegration in the Antarctic Peninsula." *Antarctic Peninsula Climate Variability: Historical and Paleoenvironmental Perspectives, Antarct. Res. Ser*, *79*, 79-92.

Scambos, T. A., Bell, R. E., Alley, R. B., Anandakrishnan, S., Bromwich, D., Brunt, K., Christianson, K., Creyts, T., Das, S. & DeConto, R. (2017). "How much, how fast?: A science review and outlook for research on the instability of Antarctica's Thwaites Glacier in the 21st century." *Global and Planetary Change*, *153*, 16-34.

Scambos, T. A., Bohlander, J. A., Shuman, C. A. & Skvarca, P. (2004). "Glacier acceleration and thinning after ice shelf collapse in the Larsen B embayment, Antarctica." *Geophys. Res. Lett.*, *31*.

Shepherd, A., Ivins, E., Rignot, E., Smith, B., van den Broeke, M., Velicogna, I., Whitehouse, P., Briggs, K., Joughin, I., Krinner, G., Nowicki, S., Payne, T., Scambos, T., Schlegel, N., Agosta, G. A. C., Ahlstrøm, A., Babonis, G., Barletta, V., Blazquez, A., Bonin, J., Csatho, B., Cullather, R., Felikson, D., Fettweis, X., Forsberg, R., Gallee, H., Gardner, A., Gilbert, L., Groh, A., Gunter, B., Hanna, E., Harig, C., Helm, V., Horvath, A., Horwath, M., Khan, S., Kjeldsen, K.

K., Konrad, H., Langen, P., Lecavalier, B., Loomis, B., Luthcke, S., McMillan, M., Melini, D., Mernild, S., Mohajerani, Y., Moore, P., Mouginot, J., Moyano, G., Muir, A., Nagler, T., Nield, G., Nilsson, J., Noel, B., Otosaka, I., Pattle, M. E., Peltier, W. R., Pie, N., Rietbroek, R., Rott, H., Sandberg-Sørensen, L., Sasgen, I., Save, H., Scheuchl, B., Schrama, E., Schröder, L., Seo, K. W., Simonsen, S., Slater, T., Spada, G., Sutterley, T., Talpe, M., Tarasov, L., van de Berg, W. J., van der Wal, W., van Wessem, M., Vishwakarma, B. D., Wiese, D., Wouters, B. & The, I. t. (2018). "Mass balance of the Antarctic Ice Sheet from 1992 to 2017." *Nature*, *558*(7709), 219-222.

Sweet, W. V., Kopp, R. E., Weaver, C. P., Obeysekera, J., Horton, R. M., Thieler, E. R. & Zervas, C. (2017). *Global and regional sea level rise scenarios for the United States.*

Tinto, K., Padman, L., Siddoway, C., Springer, S., Fricker, H., Das, I., Tontini, F. C., Porter, D., Frearson, N. & Howard, S. (2019). "Ross Ice Shelf response to climate driven by the tectonic imprint on seafloor bathymetry." *Nature Geoscience*, *1*.

USGCRP, Ed. (2018). Impacts, Risks, and Adaptation in the United States: Fourth National Climate Assessment, Volume II Washington, DC, USA, U.S. Global Change Research Program.

Velicogna, I., Sutterley, T. & Van Den Broeke, M. (2014). "Regional acceleration in ice mass loss from Greenland and Antarctica using GRACE time-variable gravity data." *Geophysical Research Letters*, *41*(22), 8130-8137.

Short Biography

I am Robin Elizabeth Bell, and I am the PGI Lamont Research Professor at Lamont-Doherty Earth Observatory of Columbia University and a member of the Earth Institute faculty. At Lamont, I direct programs in ice sheet dynamics, lead efforts to develop innovative technology and work to improve the scientific culture especially for women. I have led ten major expeditions to the polar regions, to both Greenland and Antarctica resulting in discoveries ranging from active volcanism beneath the West Antarctic Ice sheet, to large deep lakes encased by two miles of ice to hidden mountain ranges buried by ice where water under the pressure of thick ice is forced uphill. I was the first women to chair the National Academy of Science's Polar Research Board (2002-2008) where I was instrumental in launching the International Polar Year 2007-9 that brought together over 50,000 scientists from around the globe. I also co-chaired the recent National Academy report A Strategic Vision for NSF Investments in Antarctic and Southern Ocean Research (2015) that set the priorities for Antarctic science. Recently, I chaired the National Academy of Science Review of the Draft Fourth National Climate Assessment – a comprehensive undertaking by U.S. scientists and citizens documenting the impacts of climate change around the country (USGCRP 2018). Currently, I have the great honor of serving as the President of the American Geophysical Union, or AGU as we all refer to it.

Chairwoman Johnson. Thank you very much. Dr. Moon.

TESTIMONY OF DR. TWILA A. MOON, RESEARCH SCIENTIST, NATIONAL SNOW AND ICE DATA CENTER'S COOPERATIVE INSTITUTE FOR RESEARCH IN ENVIRONMENTAL SCIENCES

Dr. MOON. Chairwoman Johnson, Ranking Member Lucas, Members of the Committee, thank you for the opportunity to testify today.

Land ice loss has serious consequences within the United States and across the globe, and I'm honored to share my scientific expertise with the Committee.

Glaciers and ice sheets are Earth's water towers. Only 2.5 percent of the world's water is fresh water, and most of that fresh water is contained within glaciers and ice sheets or land ice. As Earth's water towers, glaciers are valuable sources of drinking water, irrigation water, and hydropower. But land ice is now melting at a rapid and accelerating pace, increasing risks for hundreds of millions of people who depend on them for survival and prosperity. And it is raising sea levels across the globe.

Today, land ice loss is the biggest contributor to sea-level rise. Sea-level rise can contaminate drinking water, erode coasts, over-whelm stormwater and wastewater systems, and cause increased or permanent flooding. Over just the last 25 years, average sea level around the globe has already risen 3 inches. But because sea-level rise is not evenly distributed, some areas like regions of the U.S. Gulf Coast and eastern seaboard are already dealing with more than double this amount.

The impacts we are facing today, however, may pale in comparison to the changes we could experience in the future. If we continue on our current path of high greenhouse gas emissions, it's reasonable to expect 2.5 feet or more of sea-level rise in the next 80 years. In regions of the Gulf Coast and the eastern seaboard, that number will be significantly higher.

The Greenland ice sheet, which is more than 2 miles thick in its center and covers an area the size of Texas, California, Arizona, and Nevada combined is an important player in sea-level rise. Since the early 2000s, ice loss from Greenland has increased rapidly, and Greenland is now a primary player in land ice contribution to sea-level rise.

The cause of ice loss is clear. Greenland and glaciers around the world are melting and more rapidly spilling their ice into the sea as a direct result of warming air and warming ocean water due to manmade greenhouse gas emissions. During the last 2 decades, the science community has made substantial strides in understanding Greenland ice sheet behavior and projecting future ice loss. But for any given future greenhouse gas emissions pathway, there is still a large range in projections for how much ice Greenland will lose.

Narrowing the range of future possibilities and our projections of them is possible. The United States can lead by supporting targeted research on the physical processes that control ice sheet behavior by developing systems to collect long-term observations and by fostering iterative research that connects observations and computer models. Science will also advance more quickly and better serve the public good if strong connections are fostered among scientific disciplines and between scientists and stakeholders. You can ensure this happens by increasing coordinated opportunities for interagency funding and actively funding activities that bring together scientists and decisionmakers.

Finally, I want to emphasize a critical difference in the roles of science and policy in addressing land ice loss and its impacts. Increasing scientific knowledge is essential to more accurately project what the future is likely to bring given that we are on a particular emissions pathway. But policy has the power to determine which emissions pathway we take. Embarking on a lower-emissions strategy will make a fundamental difference in how much and how quickly land ice disappears. U.S. leadership on mitigating green-house gas emissions within our lifetimes will reverberate to positively impact the world for millennia.

Thank you for giving attention to this important topic. You have the power to make a difference between a manageable future and a painful one. I look forward to supporting you with complete and accurate science and to answering your questions.

[The prepared statement of Dr. Moon follows:]

Rapid Land Ice Loss and Impacts for the United States. Statement of Twila A. Moon, Research Scientist, National Snow and Ice Data Center, Cooperative Institute for Research in Environmental Sciences, University of Colorado, Boulder before the Committee on Science, Space, and Technology, U.S. House of Representatives for the hearing Earth's Thermometers: Glacial and Ice Sheet Melt in a Changing Climate, 11 July 2019.

Chairwoman Johnson, Ranking Member Lucas, and members of the House Committee on Science, Space, and Technology, thank you for the opportunity to testify on the critical issue of rapid global land ice loss, its implications, and the challenges and opportunities for moving forward. I am heartened to see the Committee taking up this topic; sea-level rise and other impacts from land ice loss have serious consequences within the United States and across the globe. I am honored to inform the Committee's knowledge and actions via my testimony. I am a Research

Scientist at the National Snow and Ice Data Center[49] and the Cooperative Institute for Research in Environmental Sciences[50] at the University of Colorado Boulder, and my testimony today is as an expert in glacier and ice sheet science. My comments represent the views of a scientific expert, not those of the University of Colorado. Since I aim to synthesize science information within this testimony, I have provided academic paper references sparingly, but would be happy to provide additional resources on any topics or statements contained herein.

The Importance of Land Ice

The title of this hearing refers to glaciers and ice sheets - Earth's land ice - as "Earth's thermometers". It may be more accurate to describe glaciers and ice sheets as Earth's water towers. Only 2.5% of the world's water is fresh water, and more than two thirds of this fresh water is contained in glaciers and ice sheets[51]. These global water towers are critical as sources of fresh water and as long-term reservoirs to store it.

Glaciers around the World Provide Water for Drinking, Irrigation, Energy, and Other Uses

And because warm, sunny weather produces ice melt, glacier-produced fresh water is often an abundant and vital resource during dry seasons or drought. For example, roughly 800 million people in Asia, including less geopolitically stable regions across Pakistan, Afghanistan, and India, depend on glacier melt as an important water source. During average years, land ice in the region provides a small percentage of the total water supply, but in drought years glacier melt becomes much more important, sustaining millions of people and the economies they depend on. Glacier

[49] For more information: nsidc.org

[50] For more information: cires.colorado.edu

[51] Shiklomanov, I. (1993), World fresh water resources in 'Water in crisis: A guide to the world's fresh water resources'. P. H. Gleick (ed.).

melt is important in the production of hydropower as well[52]. However, ice in the region is melting more quickly than it is being resupplied. This provides a temporary bump in water availability, but future water and energy shortages due to glacier loss are certain.

As a source of cold water, glacier melt helps to regulate stream temperatures and sustain ecosystems. In Glacier National Park, for example, glacier melt is critical to maintaining the cooler water temperatures needed for native fish species and for a cascade of invertebrates that form a fundamental portion of the food web[53]. Glacier melt also helps to cycle nutrients by eroding rocks and transporting sediments underneath the ice to river and ocean ecosystems. In Alaska, the roughly $1B salmon fishing industry depends strongly on the nutrients and water properties of the Alaska Coastal Current. Roughly half of the Alaska Coastal Current waters come from seasonal glacier melt[54]. But Alaska is losing ice; the region is the second largest contributor to ice loss in the Arctic (after Greenland). Ongoing ice loss is likely to affect the future of the salmon industry and the many people who depend on it.

As the world's water towers, Earth's glaciers and ice sheets protect our coastal communities and economies from a myriad of impacts related to sea-level rise, including coastal erosion, frequent or permanent flooding, saltwater inundation of fresh water resources, disruption of storm water systems, and destruction of infrastructure including schools, homes, superfund sites, and military bases[55]. These impacts are already evident within the U.S. and across the globe as a result of almost 3 inches of sea-level rise during the last 25 years. But additional sea-level rise from ice

---

[52] Case study details: Pritchard, H. D. (2019), Asia's shrinking glaciers protect large populations from drought stress, *Nature*, 1–20, doi:10.1038/s41586-019-1240-1.

[53] For example: Clark, A. M., J. T. Harper, and D. B. Fagre (2018), Glacier-Derived August Runoff in Northwest Montana, *Arct Antarct Alp Res*, *47*(1), 1–16, doi:10.1657/AAAR0014-033.

[54] Case study details: O'Neel, S. et al. (2015), Icefield-to-Ocean Linkages across the Northern Pacific Coastal Temperate Rainforest Ecosystem, *BioScience*, *65*(5), 499–512, doi:10.1093/biosci/biv027. For a more accessible summary: https://alaska.usgs.gov/products/pubs/2014/2014_Timm_ONeel_etal_ACCC_website_factsheet.pdf

[55] For more information on the military and sea level rise: climateandsecurity.org/militaryexpertpanel

contained within Greenland, Antarctica, and the rest of the Earth's glaciers will likely dwarf what is currently being experienced. Midrange projections for additional sea level rise by 2100 are 17 inches for IPCC RCP2.6, 21 inches for RCP4.5, and 29 inches for RCP8.5, often referred to as the 'business as usual' pathway[56]. Note that these projections do not include mechanisms for potential accelerated ice loss that scientists are currently studying, particularly in Antarctica.

Recent & Rapid Global Land Ice Loss

Creation of the world-leading American economy has come about, in part, because of a stable and reliable climate. Development has depended upon and reinforced well defined coastal margins, allowing 39% of the U.S. population to thrive and prosper in shoreline counties[57], driving the U.S. economy and providing stable locations to base infrastructure for shipping, military activities, and other functions vital in today's connected global economy. This stability, however, is at risk. Since 1993, global average sea-level has already risen almost 3 inches. But sea-level rise is not evenly distributed around the world, and some communities have already experienced much higher sea-level rise than the global average. In the United States, for example, coastal erosion related to sea-level rise is displacing communities in Louisiana, drinking water problems are affecting California, and there is increased regular flooding across the Gulf Coast and Eastern Seaboard[58] (Figure 3). Similar impacts are being felt around the world.

---

[56] IPCC (2013), *Climate Change 2013: The Physical Science Basis. Contribution of Working Group I to the Fifth Assessment Report of the Intergovernmental Panel on Climate Change*, edited by T. F. Stocker, D. Qin, G.-K. Plattner, M. Tignor, S. K. Allen, J. Boschung, A. Nauels, Y. Xia, V. Bex, and P. M. Midgley, Cambridge University Press, Cambridge, United Kingdom and New York, NY, USA. p. 1180.

[57] oceanservice.noaa.gov/facts/population.html

[58] Moon, T. A. et al. (2019), The expanding footprint of rapid Arctic change, *Earth's Future*, 1–13, doi:10.1029/2018EF001088. And references therein.

Figure 3. Flooding and storm damage in the US are connected to rapid global land ice loss. a) South Ponte Vedra Beach, Florida, b) Black Creek area near Jacksonville, Florida, c) Charleston, South Carolina, and d) Lumberton, North Carolina. Images: James Balog/Earth Vision Institute[59].

As noted, sea-level rise is not the same everywhere. Local sea-level rise is due to the combined effects of local vertical land motion (some land is sinking, some is rising, depending on various factors), ocean and atmospheric currents moving ocean waters towards or away from the coast, additions of groundwater to the ocean, expansion of ocean water as it warms, and added ocean water due to loss of land ice. Today, the largest of these contributors is land ice loss. Land ice loss has increased quickly and across the globe since the mid-20th Century, with contemporary rates - and extents - of ice loss unprecedented over human history.

Satellites are a critical tool for studying contemporary changes in ice mass. The NASA GRACE and GRACE-FO (Gravity Recovery and Climate Experiment / -Follow On) satellites directly measure ice loss via changes created in the gravitational pull of the ice mass. The NASA ICESat and ICESat-2 (Ice, Cloud, and land Elevation) satellites use lasers reflected off of the Earth surface to measure the changing surface elevations of glaciers and ice sheets. Data from the joint NASA/USGS Landsat satellites are used to measure the surface speed of land ice and to

[59] Figure from: Moon, T. A. et al. (2019), The expanding footprint of rapid Arctic change, Earth's Future, 1–13, doi:10.1029/2018EF001088.

map its extent. Combining these speed measurements with observed and modeled precipitation and ice thickness information tells us how much snow is accumulated on an ice sheet versus how much ice is lost from melting and calving of icebergs into the ocean. Subtracting the losses from the gains tells us how much the ice mass has changed. These techniques alone provide three methods to independently and accurately measure ice mass changes, and results from these techniques give a consistent picture of rapid worldwide ice loss that has accelerated. This conclusion is supported by other data, including long-term measurements of glacier advance/retreat from committed individuals and institutions, examination of historical photographs, and deduction from geologic records. It is now unequivocal that contemporary ice loss is a direct result of warming air and warming ocean water due to human-caused climate change.

Role of the Greenland Ice Sheet

The Greenland Ice Sheet reaches more than two miles thick at its center, with hundreds of fast-moving outlet glaciers along its edges, which act as conveyor belts to move ice from the ice sheet interior to the ocean. Smaller glaciers and ice caps (regions of land ice that are smaller than ice sheets, but include areas of ice connecting multiple glaciers) began losing ice due to climate change earlier in the 20th century. The Greenland Ice Sheet was mostly in balance through the 1980s, but that changed in the late 1990s[60]. During the 21st century, Greenland has lost ice at an increasing tempo.

Greenland ice is being lost through melt on the ice sheet surface and where the ice contacts the oceans along its edge. Ice is also being lost as large and small icebergs that break or 'calve', into the ocean. While glaciers and ice caps continued to add the most water to the world's ocean up into the early 21st century, ice loss from the Greenland Ice Sheet has

---

[60] For example: Mouginot, J. et al. (2019), Forty-six years of Greenland Ice Sheet mass balance from 1972 to 2018, *Proc National Acad Sciences*, doi:10.7280/D1MM37.

recently reached the same level[61], with a reservoir of potential future sea-level rise that eclipses small glaciers and ice caps.

There are several indicators of current rapid ice loss in Greenland. Widespread surface thinning and ice edge retreat is observed around the entire ice sheet, exposing new land and ocean water. While personally conducting ship-based field work in northwestern Greenland in August 2018, our ship's navigation map did not show the elevation of the ocean seafloor underneath us. Instead, because the ice sheet edge in the area had retreated almost two miles since 2000, the map showed our ship ostensibly motoring across the ice sheet itself (Figure 4)! Individual glacier retreat of multiple miles since 2000 is common[62]. Along with adding ice to the ocean, the speed at which the Greenland coastal environment is changing may increase risk across valuable industries like tourism and resource extraction.

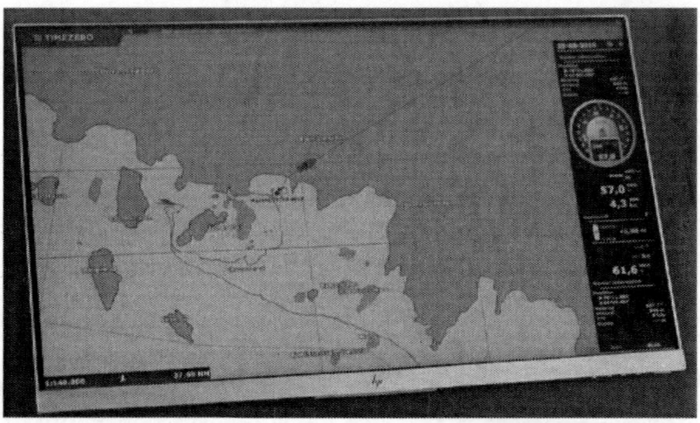

Figure 4. Rapid retreat of the Greenland ice sheet edge is changing coastal boundaries in Greenland, and in the U.S., where ice loss creates rising sea levels. Here, a science research ship appears to be traveling across the ice sheet itself because navigational maps are not keeping pace with the fast ice edge retreat. Image: Twila Moon.

---

[61] For example: Chen, X., X. Zhang, J. A. Church, C. S. Watson, M. A. King, D. Monselesan, B. Legresy, and C. Harig (2017), The increasing rate of global mean sea-level rise during 1993–2014, *Nature Climate change*, 7(7), 492–495, doi:10.1038/nclimate3325.

[62] The NASA MEaSUREs (Making Earth System Data Records for Use in Research Environments) Program provides valuable data on a variety of Greenland Ice Sheet metrics, including glacier advance/retreat. To overview NASA MEaSUREs data at NSIDC: nsidc.org/data/measures.

The rapid ice loss in Greenland is akin to the rapid melt you see in your kitchen when placing an ice cube in a water glass or leaving it lying on the countertop. The ice sheet is melting in response to warming air temperatures and ocean temperatures that are warming at depths that matter to ice (~600-1300 ft). This handy analogy has its limits, however, as the Greenland Ice Sheet system is much more complex than a household block of ice and understanding its behavior requires continued research:

- Depending on the season, the ice sheet surface is a mix of bright, reflective new snow; darker and often dirty bare glacier ice; and very dark melt lakes fed by an extensive system of surface streams. The character of the surface, which changes from region to region and over time, helps to determine how much energy, or heat, the ice sheet absorbs to melt more ice or reflects, minimizing the amount of melt. Research indicates that there will be more surface melt in the future, but also that year-to-year variations in the quantity of surface melt will span a wider range than in the past. How will the ice sheet surface transform in coming decades and respond to year-to-year variations in weather?
- Underneath the ice sheet is an entire landscape of mountains and valleys. In some areas, the ice sheet is frozen to the land, or 'bed', beneath it while in other areas liquid water and water-saturated sediments sit between the ice and the bed, affecting how easily the ice moves (or slides) over the land. Understanding the properties, and even the shape, of the bed beneath the ice is challenging, but it also vital to determining how the ice sheet moves. Will small topographic features not yet mapped speed up or slow down ice loss? How will the ice sheet be affected by increasingly large amounts of meltwater flowing underneath it?
- Some areas of the ice sheet edge end on land, but there are also hundreds of fast- moving outlet glaciers that connect directly to the ocean, each with their own unique width, depth, slope, and flow path. These areas where the ice sheet interacts with the ocean are hot spots for activity. This is where melt water discharges from

underneath the glaciers into the ocean, warm ocean water at depth contacts the glacier edge, and icebergs calve into the ocean. The ocean plays a substantial role in determining how quickly an outlet glacier can retreat and flush its ice into the sea. If the elevation of the land underneath the glacier drops further inland, retreat can create a cascade of glacier speedup and thinning that drives further retreat and may create a self-reinforcing cycle of ice loss. How will interactions between the ice sheet and ocean change the speed of ice loss? How will the fresh water from Greenland transform the ocean water, ocean currents, and marine ecosystems?

Ice loss from Greenland has impacts in addition to sea-level rise. The additional fresh water added to the ocean is changing the temperature and salinity of ocean water, and also changing the quantity and cycling of important nutrients[63]. It is likely that these changes are influencing plants and animals around Greenland, including in areas supporting important commercial fisheries. The fresher and colder water that Greenland adds to the ocean may also alter ocean currents in the North Atlantic region. North Atlantic ocean currents help to determine the climate in North America and Europe, and research is ongoing to understand how quickly and to what degree ocean currents may change, and the contributing role of Greenland Ice Sheet melt.

The past two decades have seen strong advances in characterizing, understanding, and predicting changes in the Greenland Ice Sheet. In the early 1990s our knowledge of the large ice sheets was so limited that the science community did not consider them major players in climate change on decadal timescales. Today, we understand that ice sheets are major players, and changes in Greenland are already impacting U.S. communities and the U.S. economy. The Greenland Ice Sheet, however, remains a remote and difficult place to study, and increasing our ability to predict what will happen to the Greenland Ice Sheet in coming decades will

---

[63] For example: Cape, M. R., F. Straneo, N. Beaird, R. M. Bundy, and M. A. Charette (2018), Nutrient release to oceans from buoyancy-driven upwelling at Greenland tidewater glaciers, *Nat Geosci*, 1–8, doi:10.1038/s41561-018-0268-4.

require investment in observing systems, understanding the physical processes controlling ice sheet change, and building scientific capacity and stakeholder connections.

Understanding the Future of Greenland Ice Loss: The Powerful Impact of Science

Satellite observations in the early 2000s revealed that the Greenland Ice Sheet was starting to respond to climate change. Since then, the science community has made substantial progress in understanding the behavior of the ice sheet and projecting future changes, and watched as the initial changes became much larger and more rapid. The United States has played a key role in these advances, supported via investment in satellites, in airplane- and ship-based surveys, and through advances in modeling. It is clear that while there may be short-term (less than a decade) departures from the current trend of ice loss, there is no expectation of a long-term (multi-decadal) re-stabilization under current greenhouse gas emission rates.

There remains, however, a large range in projections of ice loss for each individual future greenhouse gas emissions scenario. This spread is due to remaining limitations in our knowledge of ice sheet physics and ice sheet - atmosphere - ocean interaction, and the challenges in fully incorporating newly discovered physical relationships into computer models. The spread in the range of projected ice loss adds to the difficulty in pinpointing how much sea-level rise our nation, states, counties, and cities should plan for when making infrastructure investments. Knowing when two feet of sea-level rise will arrive to the Texas coast, for example, is essential for planning for improvements to coastal infrastructure or adjusting regional flood mitigation and planning.

Glaciology (the study of land ice) and the general study of the Earth's cryosphere (all things frozen) is a fairly young field of research, given the historic difficulty of access and vast expanse of land ice. Current resources support only a limited number of cryosphere scientists. Yet, we are trying to understand one of the most rapidly changing elements on our planet. The Arctic is warming at more than double the speed of the global average.

Additional glaciological research is necessary if we want to narrow the range of future ice loss projections and inform our Nation and the world about the ramifications for sea-level rise, water resources, etc., and the associated implications for decision-making. The United States can lead this effort via focused support to: 1) conduct research in challenging locations and understand the physics of Earth's land ice and its interactions with the atmosphere and ocean, 2) foster iterative research to connect observations and computer models to improve projections, and 3) create new and stronger connections across scientific disciplines and with decision makers through knowledge co-production so that science can advance quickly while serving societal needs.

Critical Research in Challenging Locations

Harsh weather, remote locations, large distances, and sparse infrastructure conspire to make studying Earth's land ice challenging and expensive. Yet understanding how massive ice sheets behave and how physical processes work at critical interfaces, like the ice-ocean boundary, is imperative for improving our projections of ice loss and impacts such as sea- level rise. Strengthening funding for focused research on processes controlling ice sheet behavior is vital. Through these efforts, we will increase understanding of ice-ocean interactions, ice sheet surface properties, ice sheet hydrology, and iceberg calving to improve future projections. Increased support must also include developing long-term observations that span all seasons over multiple decades. The U.S. Arctic Observing Network (AON) and international Sustaining Arctic Observing Networks (SAON) can play a valuable role in supporting these observations for the Arctic.

Connecting Current Observations to Future Projections

Observations fundamentally inform scientific knowledge about land ice processes, and are critical in developing better computer models to help scientists explore the causes of observed changes and project what will happen in coming decades. Not all physics can be seamlessly included in computer models - some processes must be captured using approximations, or 'parameterizations', that are created through observational analysis. Consistent and efficient information sharing between modeling research and observational research can fast track scientific understanding. Achieving this will require sustained support and encouragement of science that integrates these tools.

Building Collaborations to Lead

Understanding the expansive and compounding impacts of land ice loss require that new collaborations be created across different research disciplines, including areas of Earth science, social sciences, and infrastructure and economy research. Old approaches that silo science within specific disciplines must be broken down. The United States has the opportunity to be a world leader in creating strong collaborations between scientists and stakeholders. These collaborations need to be fostered via long-term, repeat interactions so that the United States can form and champion co-production of scientific knowledge and rapid information sharing. This should include support to coordinate science and stakeholder communities so that decisions at all levels are informed by the most robust science, not simply the latest headline. U.S. federal action that can support these advances by increasing coordinated opportunities for interagency funding and actively funding activities that integrate scientists with decision makers and planners. In my experience, the most effective stakeholder-scientist partnerships fully support participation across all groups.

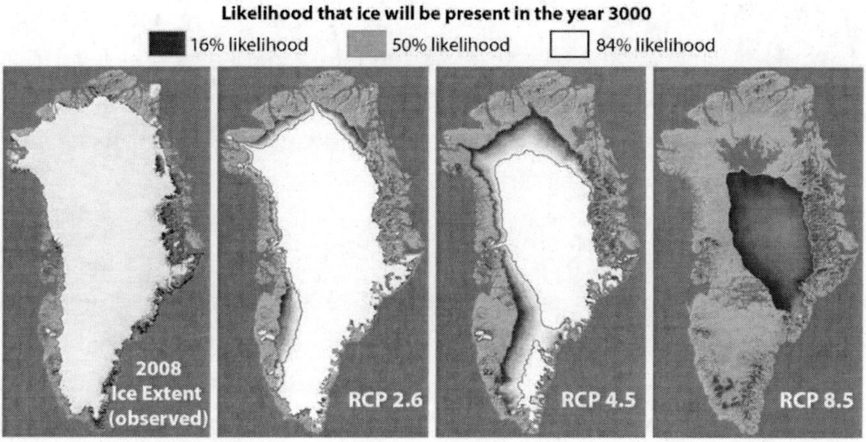

Figure 5. Computer simulations show that greenhouse gas emissions pathways followed this century will result in dramatically different Greenland Ice Sheet extents in the year 3000. All ice lost will contribute to global sea-level rise[64].

International collaboration is also key. Land ice loss and sea-level rise are not just U.S. concerns, and coordinating research in the U.S. with international efforts will serve all. The current 17-nation MOSAiC mission[65] and the U.S. - U.K. International Thwaites Glacier Collaboration[66] are excellent examples. Research in Greenland is also enhanced via strong partnerships with Greenland, Denmark, and other Arctic nations.

Creating the Future of Greenland Ice Loss: The Powerful Impact of Mitigation

The impacts of land ice loss are visible today. More difficult to comprehend is that our climate has not caught up with the greenhouse

[64] Figure modified from: Aschwanden, A., M. A. Fahnestock, M. Truffer, D. J. Brinkerhoff, R. Hock, C. Khroulev, R. Mottram, and S. A. Khan (2019), Contribution of the Greenland Ice Sheet to sea level over the next millennium, Sci Adv, 5(6), eaav9396, doi:10.1126/sciadv.aav9396.

[65] For more information: mosaic-expedition.org

[66] For more information: thwaitesglacier.org

gases that have already been added to the atmosphere, meaning that more warming and additional ice loss is certain. Continued sea- level rise over the next several decades is guaranteed and will require adaptation. If greenhouse gas emissions are reduced, however, sea-level rise can still be kept to a more manageable level.

The full power of mitigation to create the future of Earth's land ice is underappreciated. U.S. leaders, including Congress, must think beyond the short time scales of politics and realize the power they now hold for fundamentally creating the world our grandchildren and their grandchildren will live in. Recent research using a cutting-edge Greenland Ice Sheet computer simulation tells us that choosing to lead the world on a path of reduced greenhouse gas emissions can be the difference between maintaining most of the Greenland Ice Sheet or allowing it to melt completely within the next 1000 years (Figure 3).

Let me be clear here about how I see the various roles of science, policy, and leadership. Increasing scientific knowledge is essential to more accurately project what the future is likely to bring given that we are on a particular greenhouse gas emissions pathway. But policy has the power to determine which emissions pathway we take. Embarking on a lower emissions strategy will make a fundamental difference in how quickly - and how much - land ice disappears. U.S. leadership on mitigation within our lifetimes will reverberate to positively impact the world for millennia.

Communicating Scientific Urgency

Challenges remain for communicating with decision makers, planners, and the public about land ice loss and its impacts, including within the United States. The language of science can be difficult to understand. For example, the public meaning of 'positive feedback' is a good response or praise, while the scientific meaning of 'positive feedback' is a self-reinforcing cycle, often with negative consequences[67]. This disconnect is a

---

[67] Somerville and Hassol, Physics Today 64, 10, 48 (2011); https://doi.org/10.1063/PT.3.1296.

problem. Fostering interactive and sustained connections between scientists and decision making and public communities via intentional funding can create a shared language for clear understanding and informed planning.

It is also critical that decision makers and the public understand that land ice loss at the poles has strong impacts on low and mid-latitude countries, including the United States. These impacts are also not confined to coastal regions. For example, if limited federal resources are being used to respond to sea-level rise in coastal states, interior states may find it more difficult to secure resources to address other priorities or emergencies.

That mitigation today will not immediately stop ice loss and sea-level rise poses a challenge for maintaining momentum on mitigation while at the same time adapting to change. But strong mitigation today can reduce the worst outcomes of ice loss. America cannot only lead the world in understanding and projecting ice loss. America can also lead on actually determining the future of land ice and the extent to which ice loss will impact our citizens and the citizens of the world.

Thank you for the opportunity to testify today on glacier and ice sheet melt in Greenland and across the globe. I look forward to answering your questions and ensuring that the Committee has the most complete and up-to-date information on the science of ice loss and its impacts in the United States and around the world.

Short Biography

Research Scientist, National Snow and Ice Data Center
Cooperative Institute for Research in Environmental Sciences
University of Colorado Boulder

Dr. Twila Moon is a Research Scientist at the National Snow and Ice Data Center, part of the University of Colorado Boulder's Cooperative Institute for Research in Environmental Sciences, a world leader in Earth science. Dr. Moon's expertise is in contemporary changes in glaciers and ice sheets, and the connections among ice, climate, ocean, and ecosystem. While she has conducted work across the globe, her primary focus is on the Greenland Ice Sheet and the Arctic.

Dr. Moon's research uses tools ranging from satellite remote sensing to field work to computer simulations. Her work has been published in high-impact journals such as *Science* and *Nature*, and received extensive media coverage; for example, National Public Radio, the Associated Press, and the BBC. She is an accomplished science communicator and is leading efforts to improve science and knowledge co-production between scientists and stakeholders.

Dr. Moon has a BS in Geological and Environmental Sciences (Stanford) and an MS and PhD in Earth and Space Sciences (University of Washington). Subsequently, she was a U.S. National Science Foundation Postdoctoral Fellow (Ocean Sciences Division) at the University of Oregon and a Cooperative Institute for Research in Environmental Sciences Postdoctoral Fellow at the University of Colorado Boulder. She then held the position of Lecturer in Cryospheric Sciences at the University of Bristol, England, before returning to the National Snow and Ice Data Center. She serves on the Acting Committee for the Greenland Ice Sheet - Ocean Interactions Science Network. Additional information on her work is available at www.changingice.com.

Chairwoman Johnson. Thank you very much. Dr. Wolken.

TESTIMONY OF DR. GABRIEL J. WOLKEN, RESEARCH SCIENTIST AND MANAGER, CLIMATE AND CRYOSPHERE HAZARDS PROGRAM, DIVISION OF GEOLOGICAL & GEOPHYSICAL SURVEYS, ALASKA DEPARTMENT OF NATURAL RESOURCES

Dr. Wolken. Good morning. Chairwoman Johnson, Ranking Member Lucas, staff, and Members of the Committee on Science, Space, and

Technology, thank you very much for the invitation to come speak to you today.

As a citizen, I'm very pleased to be here. And I congratulate you on selecting this topic to consider. As a scientist, it means very much to me to be here to speak to you about evidence-based decision making, the data that we have to talk to you about today on glaciers and ice sheet change.

I live in Alaska, and Alaskans are very in touch with their surroundings. The cryosphere is that place on Earth where water is in its solid form, so snow, ice, and permafrost. Recently, while doing some fieldwork near Valdez, Alaska, it looks much like what you're seeing today. And so Valdez is in a fjord. It used to be covered by ice. Now the ice is melting quickly.

Upon completing a bathymetric survey or mapping the lake surface below the water near Valdez glacier, we were at the shoreline and reviewing our data and very happy about what we discovered because now we can start to find out how much water in the lake has contributed to the melting of the glacier that terminates into it. A woman and her dog named Elvis, a slobbering basset hound, came up to us and she says, what are you doing? And I said, well, we're trying to find out how deep the lake is. She said it's 600 feet deep. We looked at each other and said you're absolutely right. We just used $25,000 in equipment to figure that out. What did you do? She and her friend went out in a canoe and lowered a rope. And they discovered that the rope wasn't long enough, so they paddled back to shore and grabbed the rope. And then they tied the extra rope onto it, lowered it down, and they discovered that it was 600 feet deep near the glacier.

Now, she is a Valdez resident for 30 years. She said this glacier is melting faster than anything I've seen in the area. Where does all this water go? Well, the answer to that is in the oceans. And so Alaskans are keenly aware of their environment. They're keenly aware of the changes.

This same woman lives in an area where outburst floods impact her house every single year. The glacier releases tremendous amounts of water, rips out the dike, challenges the bridge, and gives them an opportunity to see the power of change. So the cryosphere is changing in Alaska, and glaciers are a part of that. It's very important for us to understand what is happening.

In Alaska we have a very large State. It's one-fifth the size of the rest of the United States. It's huge. We have thousands and thousands of glaciers. We know changes physically on three of those glaciers. We have mass balance data that began back in 1966. And with those data, we are able to understand how glacier change is happening over long-term. That is incredibly valuable to us.

So most of the information that we have today is built on the shoulders of giants and the data that they were able to start collecting a long time ago. It's important that we start that process now. So collecting data now in various places in state means that we can evaluate and quantify the amount of change that we have between now and whenever we're worried about the change. We do this so that we can build better computer simulations so that we can plan.

As policymakers and decisionmakers, it is imperative to have the right scientific information, and we cannot provide that without the money, without the funding, without the students, without the resources to be able to provide the information that is necessary for local stakeholders such as the woman in Valdez and her dog, as well as important federally mandated decisions that have to be made in this country. So evidence-based decisionmaking is what we are after in order to have sound change and be able to communicate to the local residents such as those in Valdez and Alaska so that we can actually start planning for some of these changes. Thank you.

[The prepared statement of Dr. Wolken follows:]

The Impacts of Glacier Change in Alaska, Testimony of Gabriel J. Wolken, Alaska Division of Geological & Geophysical Surveys and University of Alaska Fairbanks For the hearing entitled *Earth's Thermometers: Glacial and Ice Sheet Melt in a Changing Climate* Before the Committee on Science, Space, and Technology United States, House of Representatives Room 2318, Rayburn House Office Building, 11 July 2019

Chairwoman Johnson, Ranking Member Lucas, and members of the House Committee on Science, Space, and Technology, thank you for the opportunity to testify on the critical issue of glacier and ice sheet melt. As a citizen of this country I am grateful that you consider this issue relevant and important, and as a scientist I am honored to be in a position to provide evidence-based testimony to the committee to aid your decision making process. I am a Research Scientist and Manager of the Climate and Cryosphere Hazards Program at the Alaska Division of Geological & Geophysical Surveys and a Research Assistant Professor in the International Arctic Research Center at the University of Alaska Fairbanks. I provide testimony today as an expert in glacier science. My comments represent my views as a scientist and private citizen, and not those of the State of Alaska or the University of Alaska Fairbanks.

Key Points

- Glaciers in Alaska are in steep decline, and are among the fastest-melting glaciers on Earth.
- The rapid melting of glaciers in Alaska is a result of rising air temperatures associated with global climate change.
- Glaciers in Alaska are projected to continue to melt and contribute to global sea level rise.
- Hazards associated with rapid glacier retreat are impacting infrastructure, and threatening public safety and resource security.
- Narrowing the range in mass loss estimates requires a comprehensive and coordinated multi-agency collaboration, and a sustained long-term funding structure.

Glaciers and Ice Sheets

A glacier is a large mound of ice, snow, water, rock and sediment that forms on land and flows outward or down slope under its own weight. A glacier forms when snow that has accumulated doesn't melt and is buried by subsequent years of snow with the same fate. As more snow is added to the mound over many years, the snow near the bottom gets compressed and forms ice. In Alaska, glaciers form in high- mountain areas where they are constrained by topography, and flow down to lower elevation.

Glaciers gain mass by snow accumulation and lose mass by surface melt and runoff, and by iceberg separation and submarine melting at the glacier-water interface. During periods of net accumulation, glaciers gain mass and advance, whereas during periods of net loss, glaciers melt and retreat. Similar to an accountant who balances the credits and debits from a bank account, we monitor mass gains and losses of a glacier over time by computing mass balance.

Ice sheets, located in Antarctica and Greenland are massive expanses of glacier ice and snow that span entire continents. These masses of ice subsume the terrain and flow outward from a high point. At their margins ice sheets becomes thinner, and interact with the terrain.

The Role and Significance of Glaciers and Ice Sheets

Glaciers and ice sheets are important components of the cryosphere, or the parts of earth where water is in its solid form. Glaciers and ice sheets cover about 10% of Earth's land area and represent the majority of perennial land ice. Land ice is distributed as glaciers across high-mountain regions all over the world, but the majority of this ice is concentrated near Earth's poles. There glaciers and ice sheets play a vital role as regulators of global weather and climate, reflecting radiation to cool the planet and providing long-term stability for atmospheric and oceanic currents.

Glaciers and ice sheets are also an integral part of the hydrologic cycle. Together with other forms of perennial land ice, they hold about 69% of the freshwater available on the planet. They temporarily store freshwater at high altitudes and latitudes, serving as frozen water reservoirs that help to maintain a relatively stable sea level and providing consistent water resources. Runoff from glaciers provides critical baseline flow during warm and dry periods when other water sources are unavailable. This runoff is crucial to billions of people around the world, particularly those in arid regions who rely on glacier-sourced streams for drinking water and irrigation for agriculture. Runoff from glaciers also regulates stream temperature and provides nutrients for plants, insects, fish, and other animals. Glacier runoff is also important for hydropower production in many parts of the world.

Observed Glacier Change

As ice melt outpaces snow accumulation worldwide, glacier change has become one of the most important and widely recognized indicators of climate change. Historically unprecedented mountain glacier recession has proceeded on a global scale for the last thirty years (Zemp et al., 2015). In the Arctic, glaciers and ice caps have been experiencing increasingly negative cumulative mass balances since the early 1990s (Wolken et al., 2017), and are a leading contributor to global sea level change despite their relatively small size compared to ice sheets in Antarctica and Greenland (Gardner et al., 2011, 2013; Jacob et al., 2012).

Over the last several decades, Alaska has warmed twice as fast as the rest of the United States. Glaciers in Alaska (including northwest Canada) represent about 22% of the Arctic land ice area and currently have one of the highest glacier melt rates in the world (Gardner et al., 2013; Pfeffer et al., 2014), with annual thinning rates that reach several meters (tens of feet) per year for some glaciers terminating near sea level (VanLooy et al., 2006; Larsen et al., 2007; Larsen et al., 2015). This rapid and sustained melting and retreat of glaciers in Alaska is expected to continue as a consequence of climate warming, as predicted unequivocally by all current climate models (IPCC AR5, 2013; Huss and Hock, 2018).

Impacts of Glacier and Cryosphere Change

The rapid and sustained melting of glaciers in Alaska has led to a disproportionately large contribution to global sea level rise. This trend is projected to continue with inevitable consequences including lowland flooding and the displacement of communities, increased coastal erosion, degradation of water quality and agricultural soils, and loss of ecological habitats among other effects already impacting communities in the United States and elsewhere around the world.

As high glacier mass loss rates continue, the storage capacity of snow and ice will diminish, which is projected to have significant impacts on future water resource availability, even in basins with low glacier coverage (Huss and Hock, 2018). As a result, river discharge volume and timing in seasonal river runoff will change, and the ability of glaciers to sustain this flow during warm and dry periods will be reduced or lost. In Alaska, this will continue to produce clear and profound impacts on marine and terrestrial ecosystems at multiple levels, negatively influencing key species, including salmon, as well as the billion dollar fishing industry on which people depend. These changes will also heavily impact hydropower production, tourism, socio-economics, and the livelihoods and lifestyles of many people.

Rapid changes in climate can have a major effect on the cryosphere system and cause an increase in hazards. This is because changes in climate can modify natural physical processes and increase the magnitude and frequency of certain cryosphere hazards (e.g., avalanches, floods, erosion, slope instability, glacier collapses, and glacier lake outburst floods), which if not properly addressed, may have a damaging effect on Alaska's communities and infrastructure, as well as on the security and livelihoods of Alaskans. Rapid glacier retreat in recent decades has caused new ice- dammed lakes to form in valleys that were formerly occupied by glaciers. These new lakes are inherently unstable, with many resulting in glacial lake outburst floods that threaten communities and damage infrastructure (as in Juneau and Valdez). Rising air temperature and an increase in melt for Alaska's glaciers has also caused slopes surrounding glaciers to become unstable, which has led to an increase in the magnitude and frequency of landslides, and placed Southeast Alaska at the epicenter of these catastrophic events (Coe et al., 2016).

Improving our Understanding of Future Glacier Melt

Fifty-three years ago, a small team of U.S. Geological Survey scientists ventured onto two glaciers in Alaska to collect data to help understand snowpack variations in high-mountain settings. Those measurements on the Gulkana and Wolverine glaciers became the first data entries into Alaska's Benchmark Glacier program, which has provided some of the longest continuous data for evaluating the health of glaciers in Alaska. Although many advances in glacier monitoring have been made since the beginning of the Alaska Benchmark Glacier program, the observational data from this program are still used every year and represent one of the three important tools that support modern glacier science.

Modern glacier science relies on observations, remote sensing, and modeling to measure glacier melt and simulate future changes. In Alaska, the observational record is long, but there are only three long-term, continuous records of glacier mass balance for the entire state, which currently hosts many thousands of glaciers. Because of the lack of ground-based observational data, remote sensing and computer modeling are used to broaden the scope and extend estimates of glacier melt into the future based on a range of emission scenarios. But the current range in glacier ice loss projections can be narrowed. This can happen by increasing the number and distribution of long-term, ground-based observational data, increasing the quality and coverage of remote sensing products, and improving melt models and climate data products so that they more accurately represent the variables and processes involved in glacier melt and runoff in climatically diverse and topographically complex areas of the state. To achieve this, a comprehensive and coordinated multi-agency collaboration and a sustained long-term funding structure are required.

References

Coe, J. A., Bessette-Kirton, E. K. & Geertsema, M. (2018). Increasing rock-avalanche size and mobility in Glacier Bay National Park and Preserve, Alaska detected from 1984 to 2016 Landsat imagery. *Landslides*, *15*(3), 393-407.

Gardner, A. S., Moholdt, G., Wouters, B., Wolken, G. J., Burgess, D. O., Sharp, M. J., Cogley, G. J., Braun, C. & Labine, C. (2011). Sharp acceleration of mass loss from Canadian Arctic Archipelago glaciers and ice caps. *Nature*, *473*, 357-360.

Gardner, A. S., Moholdt, G., Cogley, J. G., Wouters, B., Arendt, A. A., Wahr, J., Berthier, E., Hock, R., Pfeffer, W. T., Kaser, G., Ligtenberg, S. R. M., Bolch, T., Sharp, M. J., Hagen, J. O., van den Broeke, M. R. & Paul, F. (2013). A consensus estimate of glacier contributions to sea level rise: 2003 to 2009. *Science*, *340*, 852-857.

Huss, M. & Hock, R. (2018). Global-scale hydrological response to future glacier mass loss. *Nat. Clim. Chang*, *8*(2), 135–140 doi:10.1038/s41558-017-0049-x.

IPCC. (2013). Climate Change 2013: *The Physical Science Basis. Contribution of Working Group I to the Fifth Assessment Report of the Intergovernmental Panel on Climate Change*. Cambridge University Press, Cambridge, United Kingdom and New York, USA doi: 10.1017/CBO9781107415324.

Jacob, T., Wahr, J., Pfeffer, W. T. & Swenson, S. (2012). Recent contributions of glaciers and ice caps to sea level rise. *Nature*, *482*, 514–518.

Larsen, C. F., Motyka, R. J., Arendt, A. A., Echelmeyer, K. A. & Geissler, P. E. (2007). Glacier changes in southeast Alaska and northwest British Columbia and contribution to sea level rise. *Journal of Geophysical Research: Earth Surface*. *112*(F1), doi: 10.1029/2006JF000586.

Larsen, C., Burgess, E., Arendt, A., O'Neel, S., Johnson, A. & Kienholz, C. (2015). Surface melt dominates Alaska glacier mass balance. *Geophysical Research Letters*. doi: 10.1002/2015GL064349.

Pfeffer, W. T., et al. (2014). The Randolph Glacier Inventory: a globally complete inventory of glaciers. *Journal of Glaciology*, *60* (221), 537-552.

VanLooy, J., Forster, R. & Ford, A. (2006). Accelerating thinning of Kenai Peninsula glaciers, Alaska. *Geophysical Research Letters.*, *33*(21), doi: 10.1029/2006GL028060.

Wolken, G., Sharp, M., Andreassen, L. M., Burgess, D., Cogley, J. G., Copland, L., Kohler, J., O'Neel, S., Pelto, M., Thomson, L. & Wouters, B. (2017). Glaciers and ice caps outside Greenland (in "State of the Climate in 2016"). *Bulletin of the American Meteorological Society*, *98* (8), 140-143, doi:10.1175/2017BAMSStateoftheClimate.1.

Zemp, M., Frey, H., Gärtner-Roer, I., Nussbaumer, S., Hoelzle, M., Paul, F. & Vincent, C. (2015). Historically unprecedented global glacier decline in the early 21st century. *Journal of Glaciology*, *61*(228), 745-762. doi:10.3189/2015JoG15J017.

Short Biography

Dr. Gabriel Wolken is a research scientist and manager of the Climate and Cryosphere Hazards Program at the Alaska Division of Geological & Geophysical Surveys, and a research assistant professor at the International Arctic Research Center, University of Alaska Fairbanks, where he is a senior scientist in the Climate Adaptation Science Center.

Dr. Wolken's expertise is in snow and glacier change, and the connection these changes have to climate and natural hazards. His research uses a combination of field-based observations, remote sensing, and computer modeling to monitor, simulate, and assess the impacts of snow and ice changes on communities, infrastructure, and resources. Through his affiliations, Dr. Wolken is dedicated to fostering scientific partnerships that increase the collection of critical data and improve the co-production of scientific products that lead to informed decision- making.

Dr. Wolken has a PhD in Earth and Atmospheric Sciences from the University of Alberta, an MA from the University of Wyoming, and a BSc from Creighton University. He was a Postdoctoral Fellow in the Arctic and Alpine Research Group and lecturer at the University of Alberta. Over the course of his career, he has authored numerous papers, reports and data products, and he is a regular contributor to governmental climate reports.

Chairwoman JOHNSON. Thank you very much. Dr. Pfeffer.

TESTIMONY OF DR. W. TAD PFEFFER, FELLOW, INSTITUTE OF ARCTIC AND ALPINE RESEARCH, UNIVERSITY OF COLORADO BOULDER

Dr. PFEFFER. Along with my colleagues, I'd like to thank you all, Chairwoman Johnson, Ranking Member Lucas, all the Members of the Committee and staff.

Like my colleagues, I was pleased, surprised, jumped at the opportunity to come and talk to you today about subjects that I've spent two-thirds of my life on. I've spent a long time living on glaciers and have had a good opportunity to see the changes and study them.

As Chair Johnson mentioned, I'm a glaciologist. I've done this for 40 years, and I've had opportunities to work in landscapes that have changed dramatically over time mostly in Alaska. I work mostly on the small glaciers of the world, the 200,000 glaciers other than the Greenland and Arctic ice sheets. And I want to talk mostly about them, and I really want to come back and focus on Alaska, which is one of the hotspots in the world both literally and figuratively in sea-level rise but also in fresh water flowing into the ocean, in fires and environmental change in the coastal regions.

These small glaciers matter for a wide variety of reasons, and I also want to try to concentrate today on the reasons that have direct ties to the United States. There are a number of global issues. Water resources, water availability from the Himalayas, for example, is going to be critical for Nepal, India, Pakistan, Bhutan, places like that. They also produce significant geo-hazards of landslides, flooding, what we call outburst flooding as glaciers retreat and leave behind very unstable steep slopes. When this happens in places like Nepal, these are very unstable landscapes in the same valleys where a lot of people live. It's one of the reasons that these hazards are as great as they are in the Himalayas. It's because we've got the mountains there, glaciers changing, and also people living in that landscape. That's one of the reasons that that's not quite so much of a problem in the United States because we are not obliged to live right next door to glaciers in most places, not at all.

They also have significant environmental impacts by changing the temperature of the waters the glaciers drain into and by changing the salinity of the water. One of the effects of Alaska that we don't understand particularly well yet but we know it's there is the fact that the ice sheet or the glacier runoff from Alaska that flows into the Gulf of Alaska and the Pacific, works its way up through a gap in the Aleutian Islands and enters the Arctic basin. And it turns out that that's quite a large chunk of the fresh water entering the Arctic basin, and that fresh water influences, among other things, the extent of sea ice in the Arctic.

We don't have a good handle on how much that flow is in part because we're not making comprehensive measurements of the water flow into the Gulf of Alaska from glaciers. As my colleague Dr. Wolken mentioned, we're not monitoring the glaciers in Alaska very well. We're not really keeping track of them. So while we can see that they're melting, we can measure their height change or we could up until recently anyway, we don't have good observations of where the water is going. We don't have good gauges measuring that flow.

One of the last things that I want to come back to in my statement again, though, is sea-level rise. As Dr. Alley pointed out, the ice sheets contain virtually all of the fresh water that's locked up on land in ice. You take all the other glaciers, about 200,000 glaciers, you only get about a foot of sea-level rise out of them if you put them all into the ocean. But they're like a big bucket with a little tiny—sorry, they're like a small bucket with a big hole in it. That water is leaving the small reservoir very fast. In effect, if you look at the combined most recent measurements of where new water coming into the ocean is coming from, more than 50 percent of it is coming from these small glaciers, and the remaining smaller percentage is coming from the ice sheets.

Now, that's right now. That's in the short term. The longer term, the ice sheets are certainly going to take over. But in the short term—and this is a term, say, on the order of 30, 40, 50 years where decisionmakers, planners, policymakers really need to have the most robust information, and they need the greatest handle on uncertainties. We have to look at the entire picture, the ice sheets and the glaciers and all of their consequences of which sea-level rise is just one.

So I'll stop there for now and be happy to continue and answer your questions.

[The prepared statement of Dr. Pfeffer follows:]

Role of Global Glaciers in Environmental Change and Sea Level Rise Testimony of w. Tad Pfeffer, University of Colorado at Boulder, For the hearing entitled Earth's Thermometers: Glacial and Ice Sheet Melt in a Changing Climate Before the Committee on Science, Space and Technology, United States House of Representatives, 10:00 a.m., July 11, 2019

Personal Introduction and Background

My name is William Tad Pfeffer. I am a glaciologist employed by the University of Colorado at Boulder, where I am a Professor of Civil, Environmental, and Architectural Engineering, and a Fellow of the University's Institute of Arctic and Alpine Research (INSTAAR). I have been at UC Boulder for 31 years, and have been an active glaciological researcher for 40 years. My particular sphere of expertise is in the study of the world's "small" glaciers – meaning all of the world's ca. 200,000 glaciers exclusive of the two ice sheets covering Greenland and Antarctica. I have worked extensively in glaciological laboratory experiments, numerical modeling, and theoretical analysis, and have conducted hundreds of field expeditions over 35 years in the Continental USA, Alaska, Canadian Arctic, Svalbard, Greenland, Antarctica, the Himalayas, and Africa. I have published over 60 papers in the refereed scientific literature, including several seminal and highly-cited studies of glacier physics and of global glacier contributions to sea level rise. I served as a co-author of the 2012 National Research Council Report "Sea Level Rise for the Coasts of California, Washington, and Oregon: Past, Present, and Future." I was also a Lead Author for Chapter 13 (Sea Level Change) of the IPCC Fifth Assessment (AR5), Working Group 1, in 2013. Most recently, I have shifted my focus to science planning and policy and to the historical development of glaciological and sea level research. Starting in 2013, I was a founding editor of the Oxford University Press Handbook Series on Planning for Climate Change Hazards. I also served in 2015-16 as a National Academy of Sciences Jefferson Fellow; in this capacity, I worked at USAID in Washington DC as a senior science advisor in the Office of Energy and Infrastructure, Europe and Eurasia.

My testimony reflects my own views and scientific judgement, and does not represent the views or positions of any institution or agency, including the University of Colorado.

While the potential for rapid sea level rise from the earth's two major ice sheets tends to be the most visible sphere of snow and ice research, there are other issues of concern to glaciologists involving ice in its many forms at the earth's surface. Ranging from permafrost in Siberia and Alaska to seasonal river ice in New England, from seasonal snow in Colorado's ski country to glacier lake outburst floods (or "GLOFs") in the Himalayas, and from glacier-fed rivers in India to sea ice blocking shipping routes across the Canadian Northwest Passage, ice is a crucial element of our environment anywhere on earth where freezing occurs. As temperatures rise, melting ice mobilizes liquid water, weakens previously strong frozen materials, increases the permeability of thawing soils, speeds aqueous chemical reactions, and drives a multitude of other processes, all with the potential to dramatically alter our environment. In this testimony I will focus in problems directly involving glaciers (leaving aside some equally important issues involving permafrost, river ice, and sea ice) and briefly summarize a few of what I view as the most important outstanding environmental problems. I will also concentrate on those problems that affect the United States directly, or indirectly through economic and political reactions to environmental changes elsewhere in the world.

- Seasonal snow and glacier runoff as a water resource. Society everywhere in earth depends critically on freshwater for domestic use (cooking, cleaning, washing, etc.) as well as for agricultural irrigation, industrial use, and hydropower generation. All fresh water moving on the earth's surface starts as rain or snow, but that fraction falling at high elevations as snow will remain in place (either seasonally as snow or for many years as ice) until melting conditions at the surface allow the water to move downslope. Water stored in the mountains as snow and ice acts as a reservoir, delaying the drainage of precipitation, which may arrive in very imbalanced "wet season/dry season" cycles, until later in the season. This benefits users of the water by spreading the downstream arrival of water throughout the year and storing water as glacier ice during wet years to be released during dry years. This is a critically important benefit for agriculture anywhere (including the Western US), but no more so than in the Indian Subcontinent, where very large populations depend upon crops irrigated by runoff from the Himalayas. Recent research (Maurer et al, 2019) indicates that the glaciers of the Himalayas are experiencing losses at rates that have doubled over the past 40 years. The economic and political effects of major seasonal water shortages in India and neighboring countries – a probable consequence of continued snowpack depletion and glacier losses – could be profound and global in its indirect consequences.
- Glacier recession and geohazards. People and infrastructure living in the immediate vicinity of glaciers are exposed to natural hazards including flooding, landslides, and rockfalls, all associated with slopes destabilized by the removal of glacier ice (Richardson and Reynolds, 2000) Such risks are global in extent but are particularly concentrated in parts of the world with high population densities in mountain regions, and specifically on the south side of the

Himalayas (Bhutan, India, Nepal, Pakistan) and in the Andes on the west coast of South America (Harrison et al, 2018). These regions (along with virtually all of the earth's mountain regions) are subject to landslide and rockfall hazards, but glacier retreat dramatically magnifies these hazards. Advancing glaciers disaggregate rocks and soil at their base and margins and plow this material forward and the margins of the glacier, creating *moraines* that surround the glacier terminus and valley sides. When a glacier retreats, the moraines are left behind, and "proglacial" lakes frequently form in the enclosed depression formed between the retreating terminus and the inner side of the moraine wall. Moraines are intrinsically weak materials, being composed of an incohesive mixture of soil and rocks of many sizes; they also typically have very steep slopes. These factors all favor the incidence of slope failures and landslides, and when proglacial lakes are formed, additional hazards are created due to the easily eroded moraine dams.

Source: http://www.antarcticglaciers.org/glacier-processes/glacial-lakes/glacial-lake-outburst-floods/

Figure 6. Imja Tsho (or Imja Lake) in eastern Nepal, dammed by a terminal moraine complex. The lake has been growing rapidly since the 1960s as the Imja Glacier has retreated. Photo: Sharad Joshi, Wikimedia Commons, Edited by J. Bendle.

Glacier Runoff and Ocean Salinity

One of the most significant and rapidly changing effects of present day warming is the depletion of sea ice in the Arctic Ocean (Mueller et al, 2018). The loss of sea ice in the arctic has profound implications on local and global scales, ranging from accelerated coastal erosion on Alaska's arctic coast to alterations of the planetary energy balance as the reflectivity of the Arctic Basin drops with increasing open water. The formation and maintenance of sea ice depends on the surface energy balance of the arctic and also on the salinity of Arctic Ocean water (McPhee et al, 1998). Water entering the Arctic Basin via Bering Strait (between Alaska and Siberia) is one of the primary sources of low-salinity sea water in the Arctic Ocean. The salinity of this Pacific sea water is influenced to a significant but poorly constrained degree by the Alaskan Current (Woodgate and Aagaard, 2005), which in turn carries fresh water draining into the Gulf of Alaska from the glaciers of Alaska's south coastal mountains (Chugach & St. Elias Ranges) and interior mountains (Alaska Range, Wrangell Mountains) northward and through Unimak Pass into the Bering Sea (see Figure). The retreat of Alaska's glaciers thus has an effect – probably significant but at this point not well known – and arctic sea ice. Glacier losses in the Canadian Arctic may have a similar influence (Dimitrenko et al, 2017). The influence of Alaska's glaciers on conditions in the Arctic is not well established in part because of the absence of any comprehensive program of observations of freshwater runoff to the Gulf of Alaska. This is one of many examples of the significance of Alaska's glaciers both locally and globally, and the need to invest in research in this area.

- Glacier retreat and gravitational fingerprinting. Like a magnet drawing metal filings around its edges, the large mass of glaciers and ice sheets on land (e.g., the Greenland and Antarctic ice sheets and the glaciers in Alaska) exerts a gravitational pull that draws ocean water toward them. This creates a non-uniform sea level surface, with sea level slightly elevated adjacent to any large mass concentration and slightly lowered elsewhere. This distortion on the mean sea level surface is unique to each ice mass given its size and location, and is informally referred to as the gravitational "fingerprint" of that ice mass. When glaciers shrink, however, the magnitude of their gravitational pull decreases, and the combined result of the melting of any given ice mass is to raise the total amount of water in the ocean (raising global average sea level) and to reduce that particular "fingerprint" distortion of the sea level surface. The combined effect of the gravitational fingerprints from Alaska and, to a lesser extent, Greenland, causes relative sea level to fall all along the west coast of the United States, whereas melting from Antarctica causes a relative sea level rise. The net effect of losses from Alaska, Greenland, and Antarctica on the US west coast is reduce the local sea level rise relative to the global mean value by values ranging from ca. 40% in Washington State to ca. 15% in southern California (National Research Council, 2012). Again, Alaska's glaciers have significant effects both globally and locally.

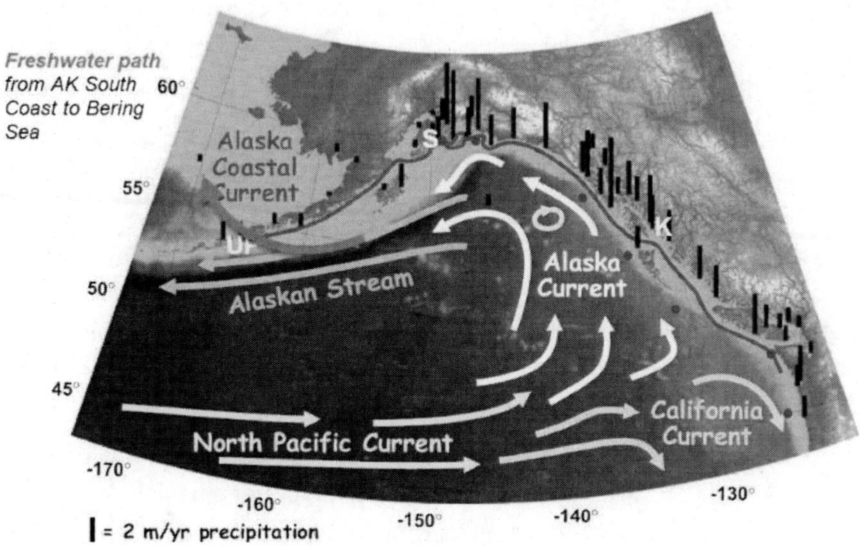

Figure 7. Alaskan coastal transport carrying glacier runoff from coastal mountains in the Bering Sea. Adapted from Weingartner et al. (2005)

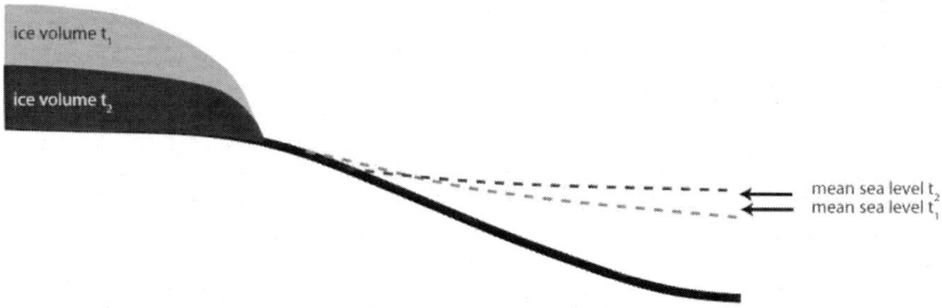

Source: National Research Council, 2012

Figure 8. Gravitational "Fingerprint" of a terrestrially-based ice mass: Globally averaged sea level increases as any terrestrially-based glacier shrinks, but *relative* sea level change is negative adjacent to the declining ice mass and positive away from it.

- Rising sea level from the earth's 200,000-plus glaciers. The total potential sea level rise from these glaciers is very small: only a bit more than 1 foot (Farinotti et al, 2018). However, the present-day *rate* of loss from the glaciers is as great as that coming from the ice sheets, and will in all likelihood continue to match the ice sheet losses for at least the next few decades, when near-term decision making requires the highest level of confidence in projections.

  Since the beginning of comprehensive global observations, virtually all glaciers on Earth have been in a state of mass loss, contributing 0.71 ± 0.08 mm yr^{-1} over the period 2003-2009 (Gardner et al., 2013) corresponding to 29±13% of the observed sea-level rise during that period. The most recent assessment of glacier losses (Zemp et al, 2019) finds a global total loss rate for the period 2006-2016 to be 0.92 ± 0.39 mm yr^{-1}. For context, the most recent ice sheet loss rate assessments show Antarctic contributing 50 ± 26 mm yr^{-1} (2008-2015) and Greenland contributing 0.77 ± 0.005 mm yr^{-1} (2007-12) and 0.53 ± 0.05 mm yr^{-1} (2012-2017). Because of their large number and small size, assessments of all 200,000+ glaciers on earth has been difficult, and the calculated aggregate loss rate has varied significantly over time, partly due to limitations in observational methods and partly due to the fact that the rates change over time. Recent research programs have benefitted from rapid developments in remote sensing, including NASA's ICESat satellite (2003 – 2009), the NASA-GFZ GRACE gravity twin satellite mission (2002-2017). Further missions, including the GRACE Follow-On (GRACE-FO), launched in May of 2018, and ICESat-2, launched in September of 2018. These mission investments have aided global glacier assessments enormously and testify to NASA's commitment to earth science generally and glacier monitoring in particular. However, remote sensing methods cannot work alone to continue accurate and validated observations of glacier change, nor can they be used in isolation to solve the numerous outstanding problems faced by modelers seeking to project future glacier behavior. Integration of field and remote sensing observations with model simulations is necessary to accurately project future trends in glacier contribution to sea level. Conventional field observations of mass balance at "benchmark" glaciers, especially those in Alaska, should remain a high priority to ensure the continuity of long-term records, some of which extend back to the 1957-58 International Geophysical Year. Ground-truthing programs are particularly important for large glacierized regions with steep gradients in environmental conditions, where the distant view of an orbiting satellite becomes a liability. Field studies at these and other sites should be expanded to include detailed observations of surface and dynamic processes. Improved parameterization of surface albedo, which controls the dominant term in the surface radiation budget, can be achieved through studies of snow and ice crystal grain sizes (Painter et al., 2009) and parameterization of the impacts of dust/black carbon (Flanner and Zender, 2006) and debris cover (Reznichenko et al., 2010) on surface melt rates. The conversion of volume to mass change in geodetic remote sensing assessments remains a large source of uncertainty (Huss, 2013) and can be informed through field measurements of near-surface densification rates. Glaciers that terminate in lakes or the ocean have the potential for rapid changes through poorly-understood calving mechanisms (Moholdt et al., 2012; Willis et al., 2018), requiring expanded observations of ice thickness, grounding line locations and lake/fiord conditions. Finally, field programs should include observations of stream discharge where possible since this provides valuable information on the integrated water balance of glacierized watersheds.

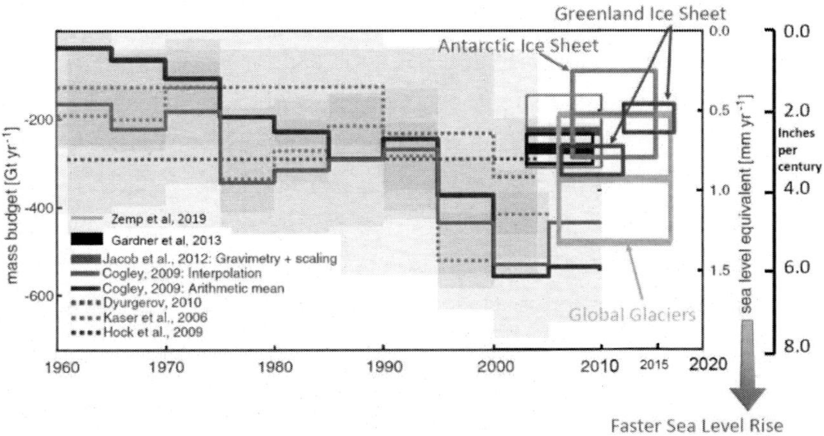

Adapted from Gardner et al, Science, 2013

Figure 9. Global glacier loss rate assessments, 1960 to the present. The most recent loss rates from the Greenland and Antarctic Ice Sheets are included for comparison.

References

Dmitrenko, I. A., Kirillov, S. A., Rudels, B., Babb, D. G., Pedersen, L. T., Rysgaard, S. & Barber, D. G. (2017). Arctic Ocean outflow and glacier-ocean interactions modify water over the Wandel Sea shelf (northeastern Greenland). *Ocean Science*, *13*(6), 1045-1060.

Farinotti, D., Huss. M., Furst, J. J., Landmann, J., Machguth, H., Maussion, F. & Pandit, A. (2019). A consensus estimate for the ice thickness distribution of all glaciers on Earth. *Nature Geoscience*, *12*(3), 168.

Flanner, M. G. & Zender, C. S. (2006). Linking snowpack microphysics and albedo evolution. *Journal of Geophysical Research: Atmospheres*, *111*(D12). Gardiner et al, 2013

Harrison, S., Kargel, J. S., Huggel, C., Reynolds, J., Shugar, D. H., Betts, R. A. & Schaub, Y. (2018). Climate change and the global pattern of moraine-dammed glacial lake outburst floods. *The Cryosphere*, *12*. Huss, 2013

Maurer et al., (2019). Maurer, J. M., Rupper, S. B. & Schaefer, J. M. (2016). Quantifying ice loss in the eastern Himalayas since 1974 using declassified spy satellite imagery. *The Cryosphere*, *10*(5), 2203-2215.

McPhee, M. G., Stanton, T. P., Morison, J. H. & Martinson, D. G. (1998). Freshening of the upper ocean in the Arctic: Is perennial sea ice disappearing?. *Geophysical Research Letters*, *25*(10), 1729-1732.

Moholdt, G., Heid, T., Benham, T. & Dowdeswell, J. A. (2012). Dynamic instability of marine terminating glacier basins of Academy of Sciences Ice Cap, Russian High Arctic. *Annals of Glaciology*, *53*(60), 193-201.

Mueller, B. L, Gillett, N. P., Monahan, A. H. & Zwiers, F. W. (2018). Attribution of Arctic Sea Ice Decline from 1953 to 2012 to Influences from Natural, Greenhouse Gas, and Anthropogenic Aerosol Forcing. *Journal of Climate*, *31*(19), 7771-7787.

Painter et al., 2009.

Richardson, S. D. & Reynolds, J. M. (2000). An overview of glacial hazards in the Himalayas. *Quaternary International*, *65*, 31-47.

Reznichenko, N., Davies, T., Shulmeister, J. & McSaveney, M. (2010). Effects of debris on ice surface melting rates: an experimental study. *Journal of Glaciology*, *56*(197), 384-394.

Weingartner, T., Aagaard, K., Woodgate, R., Danielson, S., Sasaki, Y. & Cavalieri, D. (2005). Circulation on the north central Chukchi Sea shelf. *Deep Sea Research Part II: Topical Studies in Oceanography*, *52*(24-26), 3150-3174.

Willis, M. J., Melkonian, A. K., Pritchard, M. E. & Ramage, J. M. (2012). Ice loss rates at the Northern Patagonian Icefield derived using a

decade of satellite remote sensing. *Remote Sensing of Environment, 117*, 184-198.
Woodgate and Aagaard. (2005).
Zemp, M., Huss, M., Thibert, E., Eckert, N., McNabb, R., Huber, J. & Thomson, L. (2019). Global glacier mass changes and their contributions to sea-level rise from 1961 to 2016. *Nature*, *568*(7752), 382.

Short Biography

My name is William Tad Pfeffer. I am a glaciologist employed by the University of Colorado at Boulder, where I am a Professor of Civil, Environmental, and Architectural Engineering, and a Fellow of the University's Institute of Arctic and Alpine Research (INSTAAR). I have been at UC Boulder for 31 years, and have been an active glaciological researcher for 40 years. My particular sphere of expertise is in the study of the world's "small" glaciers – meaning all of the world's ca. 200,000 glaciers exclusive of the two ice sheets covering Greenland and Antarctica. I have worked extensively in glaciological laboratory experiments, numerical modeling, and theoretical analysis, and have conducted hundreds of field expeditions over 35 years in the Continental USA, Alaska, Canadian Arctic, Svalbard, Greenland, Antarctica, the Himalayas, and Africa. I have published over 60 papers in the refereed scientific literature, including several seminal and highly-cited studies of glacier physics and of global glacier contributions to sea level rise. I served as a co-author of the 2012 National Research Council Report "Sea Level Rise for the Coasts of California, Washington, and Oregon: Past, Present, and Future." I was also a Lead Author for Chapter 13 (Sea Level Change) of the IPCC Fifth Assessment (AR5), Working Group 1, in 2013. Most recently, I have shifted my focus to science planning and policy and to the historical development of glaciological and sea level research. Starting in 2013, I was a founding editor of the Oxford University Press Handbook Series on Planning for Climate Change Hazards. I also served in 2015-16 as a National Academy of Sciences Jefferson Fellow; in this capacity, I worked at USAID in Washington DC as a senior science advisor in the Office of Energy and Infrastructure, Europe and Eurasia.

Chairwoman JOHNSON. Thank you very much. That completes the testimony of our witnesses. We'll now begin our first round of questions. And I'll yield myself 5 minutes.

While a lot of progress has been made in understanding current trends in glacier and ice sheet melt rates and predicting future changes, uncertainties still exist and the potential for tipping points. Whether major parts of the Arctic or Greenland sheets will collapse and when and how much more sea levels will rise is yet still out there.

So for all of the witnesses, as top experts in the field, what are the most pressing needs in glacier research in order to address these uncertainties? And the second question, given the differing impacts of glacier and ice sheet melts on global sea-level rise, ocean temperature and salinity, nutrient cycling and ocean currents, fisheries, and even geopolitical tensions from diminishing drinking water supplies, how can multidisciplinary research approaches help address some of the outstanding questions? And we can start and just go down the panel.

Dr. ALLEY. OK. Thank you. You raise huge and important issues. You will hear several things as we go along here which are important. I'd like to highlight the people. The students who can really solve these problems have an amazing number of calls on them. They can go to business, they can go to finance, they can do all sorts of things. They have skills that are hugely in demand. We hope that a lot of them go to business and go to finance and do useful things out there, but we would love to have a few of the best students come to us. If those students look at our world and say there isn't funding, there isn't a reliable idea that you can make a career in telling the public what's going on, they all will go elsewhere. And we don't want all of them, but we would really like a few of them. And that means funding for studentships and that means some level of telling the student if you commit to 4 years as an undergrad and maybe 7 years as a graduate student and a bit of postdoc to become a world expert on this, we will support you in doing that. And it is people. And we need a few of them to help us do this.

Chairwoman JOHNSON. Thank you.

Dr. BELL. Well, I think there are three things. As I said before, one is fostering more research, and it's research across the agencies because the U.S. has really been leading in understanding how the ice sheets are changing. And it's becoming increasingly important that research, whether it's supported by NSF, NOAA, USGS (United States Geological Survey), NASA—DOE (Department of Energy) runs a lot of ice sheet models. We have to recognize this incredible resource we have to be at the cutting-edge of who's going to know and be able to provide the answers to communities around the globe.

I echo Richard's workforce question. We really do need to broaden the number of people working on this and not just glaciologists. We need engineers. We need computer scientists. We need to, you know, recognize that this is a significant national security, a national economic issue that requires all hands on deck.

And then the third one is really to foster what NSF is now calling convergent science, science where you really bring together people from different disciplines focused on a problem so that we can address the problem. NSF just released a new priority—they're navigating the new Arctic where that was really the focus of how do we go in a changing arctic. Now getting back to Ranking Member Lucas' question, how do we have science that brings together the people who are going to look at those problems? And NSF is really trying to foster that collaborative problem solving. The word currently is convergent where you actually bring people from different disciplines who are looking to solve a problem, and that's what we absolutely must do, both within the U.S. and globally. This is a problem we cannot solve by ourselves.

Chairwoman JOHNSON. Thank you. Dr. Moon.

Dr. MOON. Yes. I'd like to emphasize first better understanding the glaciers and ice sheets themselves, the physical processes that we don't yet know. We've never been able to watch ice sheets collapse, so we can't look back into the record. And that requires going there, observing the systems, and doing process studies and then integrating that information with computer simulations.

The second thing is that any projects within this need to be coordinated with other countries. Sea-level rise, ice sheet, glacier melt are international issues, and we need international teams working on them. And anything you can do to help facilitate international collaboration I think is excellent.

And finally, your question regarding connecting different disciplines, this is a very difficult thing to do. Disciplines have been separated for decades. It's the way many elements of our academic system and our research system are built, and it requires long-term investment and an understanding that we have to create those relationships because we're taking our information about glaciers and ice sheets and we're recognizing that they're part of a connected Earth system that includes people, as well as plants, animals, and other physical components.

Chairwoman JOHNSON. Thank you.

Dr. WOLKEN. I will echo the comments of my colleagues in that we do need people. It's critical that we have people on these issues. I'll share with you just an example from one of my other projects looking at snow distribution on glaciers in alpine areas. We've gone through three people now in a critical data science position because they can make more money elsewhere. And so it's really hard to retain the people once we have them, and it's really challenging to actually recruit these people.

The other issue is, again, I'll emphasize Alaska is one-fifth the size of the rest of the United States. We have a lot of area, and we are data-poor. We have only a handful of long-term observations in the State, and we have very few long-term records. And when I mean long-term I mean beyond 12 years. So it's very challenging to work in that environment. And so what we really need is more observational information to go off of.

And, you know, we can do a tremendously better job with the science if we have data. The only way to make the models do better is to actually have data that drive the models. And so we don't have that right now. So we're doing the best we can. And one of the most important observation technologies that we had on Operation Ice Bridge in the State that measures the height of glaciers is now gone. It's been discontinued as a program, and so we can't do that anymore.

And so the other thing I'll say is that I'll echo the comments of having a mixed bag of individuals to do critical tasks, and it's important for us to have a diversity of individuals working on these really important issues.

Chairwoman JOHNSON. Thank you very much. Dr. Pfeffer.

Dr. PFEFFER. Well, after writing down my list of responses to your question, I realize that all I really have to say is I'm with them. We seem to be pretty much on the same page here, and I promise you, we didn't rehearse this in advance. It really boils down to people and support to train those people. We've got these critical questions, the research questions. One of them is the tipping points. What's actually going on in glaciers and ice sheets that causes this occasionally very anomalous behavior? We can't model that problem out of the way with computers or with our knowledge of mathematics and mechanics. We actually have to get in—literally in and under the glaciers to see what's happening.

Decades ago when I was—we don't have to run away and leave the building, do we? OK. Decades ago when I was a graduate student—and Richard will remember this, too—we had programs all through the United States with opportunities for graduate students to work on small glaciers in Alaska, in the Pacific Northwest, and the Scandinavia and Arctic, as well as in Greenland and Antarctica where we could go back a number of times while we were students, really learn what is happening on glaciers. I'd already made 12 trips to Alaska before I finished my Ph.D. That's really not happening anymore. Most departments—and we have—we've got an abundance of programs. We've got a lot of expertise out there searching for students trying to bring them in, but we don't have the support to really go places to train them. And so we are producing a lot of computer modelers, very good, and they're doing very important work, but they're waiting for this knowledge to come in for them to put into these models. And we're really falling behind on that.

Also, as Dr. Wolken mentioned, we're missing what's happening in Alaska. Operation Ice Bridge, which was our best way of tracking the loss of glaciers in Alaska, is—that's vanished for the time being. And monitoring of fresh water flowing into the ocean, we don't know what that is. So we need the answers to those questions, but to get the answers, we need people.

Chairwoman JOHNSON. Thank you very much. I'm way over. Mr. Lucas.

Mr. LUCAS. Thank you, Madam Chairwoman. Dr. Moon, let me turn to you first, and let's discuss for a moment Federal Government spends a lot of money on research, we have a lot of research dollars coming out of National Science Foundation, the various department agencies, the DOD (Department of Defense). We're doing a lot of things. Let's talk for a moment about from your experiences what kind of suggestions you have about, in a more direct way, how do we make sure that the various Federal research activities are better coordinated, integrated, as you noted?

Dr. MOON. Well, I can speak to one example that I think is a nice example for how to work on this. This is a program called IARPC, the Interagency Arctic Research Policy Committee in the Arctic. And that program brings together people who are funded and doing work supported by different agencies. They have regular webinars, they have regular meetings, and it helps people communicate on what's going on. I think having those sorts of tools and having communication between agencies at the program manager level also encourages us on the research scientist level to be able to get information that tells us about what different agencies are interested in so that we have a sense of the interest, the potential funding, and how they might be connected together.

Mr. LUCAS. Thank you. And, Dr. Bell, as I mentioned in my opening statement, I have a deep interest in weather patterns. I represent the northwest half of the great State of Oklahoma. I'm a product of my experiences, but I'm also impacted by the experiences of the people who came before me in my district. And where we are on the east side of the Rockies and the Southern Plains, my area was one of those that suffered dramatically in the Dust Bowl of the 1930s, the Great Depression.

But in addition to that horrendous drought, which had a lot of government policy and farming practices that enhanced the misery, we went through a drought in the 1890s, drought in the 1930s, drought in the

1950s, the horrendous drought at the beginning of this decade. So you understand as a farmer in the real world, where I come from, when I asked my questions, how much does the scientific community understand about what will happen in the lower latitudes weather-pattern-wise by what's going on in the glaciers? Do you see where I'm coming from here? Because glacial water changes the chemistry of the oceans, changes the temperature. I did pay some attention in my science portfolio at Oklahoma State.

Dr. BELL. Right. And one of the predictions, you know, one of the very clear predictions from the climate models is we're going to see a lot more of those extremes. We're both going to see a lot more droughts, and we're likely to see a lot more floods—many of your neighbors saw a lot more floods this year because we all know we're here in Washington in the summer and it is hot and muggy. That is because hot air holds more moisture. And in the long run, that's going to make more floods. So, you know, in terms of the direct linkage between the warming climate and the weather, that's the easiest one to think of. We are going to expect more extremes in precipitation, and we're going to expect more extremes in weather. The direct link between the changing land ice and changing climate is something we're still working on is what—we heard it from Tad Pfeffer is what will that water go into the ocean due to the ocean circulation? Where are certainly hypotheses out there that the changing sea ice have contributed to some of the extreme weather we're seeing now. You know, certainly, that's on the table. But again, it's showing how we have not decoded the weather system and the climate system on our planet, but we can see the impacts already.

Mr. LUCAS. Please, Dr. Alley.

Dr. ALLEY. Yes, Representative Lucas, I'm sorry to interrupt. As you know, the great State of Oklahoma is fantastic in educating meteorologists. And you probably also know that there really is scholarship that shows that the Federal investment in meteorology pays handsome dividends for the well-being of farmers, for the well-being of fisherpersons and others and at a level that is a huge payoff on investment. And I can give you chapter and verse if you need it.

There is great optimism now in the community that does weather forecasting that we will be able to move into that area which would give more warning to the farmers of Oklahoma, the fisherpersons of Oregon and Maine as to what's coming. We can't guarantee that, but the optimism is real, it's palpable, and it's exciting.

Dr. BELL. And I'm going to just add one thing—I just want to follow up with one thought following on Richard's is that we have invested a lot in weather forecasting. That's why we now—I'm not a farmer, I'm a sailor, so I think about hurricanes more than droughts. Sorry. But——

Mr. LUCAS. You have waterspouts, we have tornadoes.

Dr. BELL. Yes. Yes, yes, I do, and I worry about them, too. They give me goosebumps. But we've been able to narrow our understanding of where those hurricanes are. I can plan much better when I hear there's a hurricane coming than I used to, and that's because we've invested in weather research, everything from the process-based work to the numeric, and that's what we don't have for the ice sheets yet.

Mr. LUCAS. Indulge me, Chairwoman. Dr. Wolken, what are you telling the State of Alaska about how to handle the circumstances in the next decade or so?

Dr. WOLKEN. Well, that's a good question. I mean, we have some of the best climate modelers in the world at the University of Alaska Fairbanks, and they are doing a tremendous job in producing downscale climate models for Alaska. Their products are only as good as the data that they can use to train those, and they've done a tremendous job in predicting out to, say, 2100 what the climate is going to be like. From that, we can make some estimates of how glaciers are going to respond, how the cryosphere in general is going to respond. And, you know, the best tools that we can produce are available, but we need to improve those tools tremendously in order to make better predictions so people can plan.

Mr. LUCAS. Thank you, Doctor. Thank you, Chairwoman.

Chairwoman JOHNSON. Thank you very much. Ms. Bonamici.

Ms. BONAMICI. Thank you, Chairwoman Johnson and Ranking Member Lucas. But really thank you to our witness experts who are here today. We do appreciate your expertise.

I'm fortunate because I'm from the Pacific Northwest where we still have glaciers and the Cascade Range and the beautiful Wallowa mountains in eastern Oregon. And over the years the snow and ice masses have really helped delicately balance our water temperature and our ecosystems. The nutrient content, glacial melt water has provided drinking water, and the runoff helps power our communities. Tourism and outdoor recreation are really important in our State. People travel to see our streams, rivers, and lakes, which the glacial sediment makes this iconic teal color. It's a beautiful place. You should all come and visit.

But today, the glaciers that have once filled a lot of the hanging valleys and the moraines and the mountaintops across some of the most pristine regions are rapidly melting and in large part because of anthropogenic emissions.

On Mount Hood, which we can see from Portland in my home State, the Sandy Glacier Caves were once the largest glacier cave system in the lower 48 States, but now, the glacier is melting at an alarming rate. And further north of Oregon at the Columbia River basin at Glacier National Park, they're losing the geologic features that provided its namesake. In fact, when it was founded in 1910 the park had about 150 glaciers. And according to a study from Portland State University and the USGS the park is on track to lose its remaining 26 glaciers in the next few decades.

Dr. Moon, thank you for your testimony. You mentioned the role of glaciers in sustaining ecosystems, and in northwest Oregon the expedited rate of melting of glaciers could have significant consequences for our salmon and steelhead populations and threaten recreational and commercial fishing, tribe species that benefit from healthy salmon runs. And as the glaciers melt and the water flows change and the water temperatures warm in the Columbia basin, the tributaries, the fisheries are threatened. So how quickly are these larger ecosystem changes taking place? And are there potential adaptation and mitigation strategies that we in Congress can support to help at this point in time?

Dr. MOON. I would say—and you might find, too, some—that you have many problems also related to those that are being seen in Alaska and receive comments there. Certainly, we are losing those glaciers very rapidly. Just as you cited, we are seeing retreat and ice loss at rates that have never been seen in these areas. And so those fundamental changes that are happening rapidly and quickly are changing the ecosystem just as quickly. One——

Ms. BONAMICI. Right.

Dr. MOON [continuing]. Thing to consider is that in many of these places we initially see a bump in the amount of water because we're getting warmer air temperatures. We still have the glaciers there at the moment, so you actually get a bump in water availability, and we see communities also in other places in the world where they depend even more strongly on glaciers for drinking or irrigation water adjusting to an added level of water input, which then of course is eventually going to decline substantially to levels below what it was——

Ms. BONAMICI. Right.

Dr. MOON [continuing]. Previously. So they are rapid changes, and I think that there are many places where the research is not keeping up with the speed of these changes. That's true for us understanding the glaciers and ice sheets themselves and also certainly true for understanding the ecosystems that depend on it. So I think it may be a case where we are changing things that we are not even able to keep up with or see the true level of those changes.

Ms. BONAMICI. Thank you. Thank you. I appreciate that very much. And a good place for the Science Committee to get some more research funded.

Dr. Pfeffer, some of your colleagues at the University of Colorado Boulder published a study in 2017 about the effects of dissolved black carbon on glacial melting, that sooty black material that's emitted from gas and diesel engines, coal-fired power plants, and wildfires is a significant portion of particulate matter and contributes to climate change, as we know. The study found that the black carbon from the combustion of biomass and fossil fuels can enhance glacial melting as black carbon deposits on snow and ice surfaces, then the particles decrease the Earth's ability to reflect rays from the sun, so then that results in the absorption of heat and faster melting. But it's also worth noting from the testimony here today that even if anthropogenic emissions were halted immediately, we'd still see the reciprocal effects on glaciers.

So, Dr. Pfeffer, what are the most apparent gaps in the current modeling of glacial recession for various emission scenarios? And assuming that the U.S. achieved a net-zero carbon emission policy by midcentury, where should we invest more Federal resources in responding to the consequences of glacier melting?

Dr. PFEFFER. So the process that you bring up, black carbon, it's hard to see when you're actually out there. It's quite a subtle effect but very small particulate matter, which is carried into the air, and this was particularly a problem prior to the collapse of the Soviet Union because there was a lot of coal burning industry in Siberia. It's far enough north for their emissions to get trapped in this atmospheric gyre in the Arctic. That's reduced a little bit but not by a large degree. And emissions from further south still get into the Arctic basin and also elsewhere. Not all of Greenland is in the Arctic basin, for example. Southern Greenland is exposed to air masses that come off of Europe and North America, so there's a lot of mixing. And this material continues to be deposited.

I think that understanding the surface energy balance, things like if you make the surface of an ice sheet just a tiny bit darker, how much effect will that have, that understanding is pretty well in hand but we need observations. Simply knowing that it happens isn't enough. I really do think, though, that the basic needs go beyond that to simply making the observations. There are so many parts of the world that were, until recently, really in the dark. A lot of high mountain Asia, Himalayas and other ranges, that's been partially addressed by remote sensing, but again, not all of it. Some of this work just has to be done on the ground.

Ms. BONAMICI. Thank you. And I see I'm out of time, but, Chairwoman Johnson, I request unanimous consent to enter into the record this study from the University of Colorado.

Chairwoman JOHNSON. Without objection.

Ms. BONAMICI. I yield back. Thank you.

Chairwoman JOHNSON. Thank you. Mr. Brooks.

Mr. BROOKS. Thank you, Madam Chairwoman.

Is anyone on the panel not familiar with the Earth's last glacial maximum, roughly 20,000 years ago? OK. Everybody is? Good. For those in the audience who are not, by way of background, during the last glacial maximum, northern Europe was under ice, roughly 90 percent of Canada and almost all of the Continental United States of America north of Missouri and the Ohio Rivers and east of New York City under ice. According to the United States Geologic Survey, during the last glacial maximum, again, 20,000 years ago, sea levels were roughly 410 feet lower than today. Stated differently, for 20,000 years, sea levels have risen on average 2 feet per century versus the much less roughly 1 foot per century rising rate since 1993 that is reflected in Dr. Alley's written testimony.

Finally, per Zurich University of Applied Science, Earth's average temperature 20,000 years ago was 48° F versus 59° F today. That's an 11-degree increase in global temperature average over the last 20,000 year period.

So my question to each of you is, and we'll start over here with Dr. Pfeffer and move from my right to left, did human beings cause the global warming that started 20,000 years ago and continues through today or, if not, what did?

Dr. PFEFFER. So the examples from 20,000 years ago that Mr. Brooks gave us, they are excellent examples of the kind of natural variability that the Earth experiences. And there's no question that in the past, there have been changes in temperature and sea-level rise and weather patterns and climate generally as dramatic or more dramatic than what we may be experiencing in the future. And of course they weren't human-caused 20,000 years ago or in the last million years. All of these variable events have been occurring throughout the Earth's modern history.

Mr. BROOKS. Well, my first question, in your judgment, did human beings cause the global warming that began 20,000 years ago during the last glacial maximum?

Dr. PFEFFER. No. No, absolutely not. It's an example of spontaneous natural variability, one of the many ways that this whole system, whether you look at it in terms of sea-level rise or temperature, storms, can be varied.

Mr. BROOKS. Are you familiar——

Dr. PFEFFER. Natural——

Mr. BROOKS [continuing]. With the phrase snowball Earth or slush ball Earth——

Dr. PFEFFER. Oh, yes. Yes.

Mr. BROOKS [continuing]. Roughly 600 million years ago——

Dr. PFEFFER. Yes.

Mr. BROOKS [continuing]. When we were almost entirely ice or slush?

Dr. PFEFFER. Entirely natural variation.

Mr. BROOKS. Versus the Paleocene and Eocene thermal maximum of about 55 to 56 million years ago when the average temperature was roughly 73°F, which is 14 degrees warmer than what we are experiencing now.

Dr. PFEFFER. Yes.

Mr. BROOKS. If you don't mind, Dr. Wolken, let's go to you. Did human beings cause the global warming that began 20,000 years ago?

Dr. WOLKEN. No, absolutely not. That was just a product of natural variability in the climate system. Yes.

Mr. BROOKS. Dr. Moon?

Dr. MOON. Humans weren't around in nearly the numbers we are today, so we certainly weren't available to be combusting fossil fuels at the rate we are today or putting emissions into the atmosphere. You can consider we've built America in the last 243 years, and we're changing things at a much more rapid rate.

Mr. BROOKS. So you also agree then that the global warming that has occurred over the last 20,000 years, that 11°F increase in temperature was not human caused at least when it began 20,000 years ago?

Dr. MOON. So I would agree that when it began 20,000 years ago when we were coming out of the last glacial, that was not caused by humans.

Mr. BROOKS. All right.

Dr. MOON. The warming of the last hundred years most certainly was.

Mr. BROOKS. Out of curiosity, how do you explain that the sea- level rise average over the last 20,000 years has been 2 feet per century, yet we're down to 1 foot per century?

Dr. MOON. So much of our rise in sea levels that you're talking about came earlier in that 20,000 years.

Mr. BROOKS. For 6,000 or 7,000 years.

Dr. MOON. Over this last 10,000 years, we've been sitting with very stable sea levels. And those stable sea levels have allowed us to develop the coasts of the world.

Mr. BROOKS. All right. Thank you, Dr. Moon. And I only have about 30 seconds left for Dr. Bell. Dr. Bell, in your judgment, 20,000 years ago, global warming when it began, was that caused by humans?

Dr. BELL. In my judgment, the variation that we were seeing 20,000 years ago was part of the pulse of the planet. It pulses at 100,000 year glacial, interglacial. When I started graduate school, we were expecting to go into the next glacial period——

Mr. BROOKS. Yes.

Dr. BELL [continuing]. Except that we as human beings in the last hundred years—and you can see the kick up—since we invented the steam engine, you can see the temperature moving up.

Mr. BROOKS. All right. I'm out of time. Madam Chairwoman, I appreciate your indulgence. I just wish I had sufficient time to actually get into what the cause of the global warming that began 20,000 years ago was if not humans. Thank you.

Chairwoman JOHNSON. Excuse me. Go ahead, Doctor.

Dr. PFEFFER. I just wanted to respond a bit further to your question. The changes in the past have—there are two significant differences between those events and the events today. One of them is that they were triggered by natural variations, not by human agency.

And let me just give you an analogy of your house. Your house might burn down, and it might burn down for entirely natural reasons. It might be struck by lightning. But it could also burn down if you're careless and you, you know, drop a cigarette in the crack in the sofa. Both of those are triggers that result in your house burning down. The presence of one of them doesn't really say much about the other except that they both lead to the same endpoint.

The other thing is that while there were these very dramatic temperature changes and sea-level rises in the past, which were entirely natural, we weren't there to deal with them. The problem here is with people. How do we respond to environmental change?

The Earth will take care of itself. It doesn't really care what happens. It's what people do. And if this had happened, you know, a long time ago when the population of the Earth was a few hundred million, it probably wouldn't have mattered either because we could have just gotten out of the way. But as it is today with the numbers of people that we have and the infrastructure, we're very sensitive to changes of this kind. We don't handle change very well.

For example, suppose that the conditions for growing crops that exist today in California picked up and moved to North Dakota for a couple of hundred years. There are variations like that in the fairly recent geologic past that occurred. How would we deal with that? It's an entirely different world than what we were not here to experience but we know about 20,000 years ago. We're much more sensitive. We don't deal well with change, and to deal with it, we need to know a lot about it.

Mr. BROOKS. Dr. Pfeffer, thank you for your additional insight.

Chairwoman JOHNSON. Thank you very much. Mr. McNerney.

Mr. MCNERNEY. Well, I thank the Chair for calling this hearing, and I thank the witnesses. I appreciate all your testimony this morning.

While the planet is continuing to warm up and I believe we are going to blow past the 2° centigrade marker that people say is the limit of tolerability, we need to be looking at all the potential tools in the climate solutions toolbox, especially if we're to take action to prevent the collapse of the West Antarctic and Greenland ice sheets. That's why I introduced the *Geoengineering Research Evaluation Act* last Congress. It didn't pass, but just introducing that caused the National Academies of Science to explore the state of research in climate intervention strategies, as well as the need to implement a governance structure of those technologies.

Dr. Bell, given the complexity of the climate system and the risks associated with further human interference, how do you think the U.S. should approach the field of research on climate intervention?

Dr. BELL. Both the National Academy and AGU, the American Geophysical Union, have statements that say this is an issue that we must research. If done wrong, it could be terrifying. But, again, it is the same problem that we have been saying before. We don't have the sufficient workforce looking at the issue, evaluating it, and building the body of knowledge to evaluate whether or not it is a good idea.

To me, I come back to the very, very few examples of geoengineering of the ice sheets that are out there. And to give you the idea of how many groups have done it, I think two groups have put it on the table. You know, one is basically for one approach— you get a bunch of snow blowers and put more snow back on the ice sheet. The problem is it turns out if you put snow blowers on the ice sheet, it gets steeper and it flows back into the ocean. It didn't work. The other idea is to build bigger than the Panama Canal many times walls to keep the ice sheets from being attacked by the warming ocean.

These are ideas being put on the table by a small cadre of glaciologists. What this illustrates is that we need, as a species, to research this, and we need not just glaciologists, not just atmosphere scientists, but we need to bring the full suite of talent to the table to think about this because, as we address climate change, we're going to probably need to look at every tool that we have available. That's what we found when we did the building down the street.

Mr. MCNERNEY. Thank you.

Dr. BELL. We couldn't reach our goal by doing just one thing.

Mr. MCNERNEy. Dr. Alley, do the risks of abrupt change in the Arctic and Antarctica indicate that we should be serious about technological interventions such as sunlight reflection to maintain stability?

Dr. ALLEY. So I would echo what the National Academy and what the American Geophysical Union have said, which is that we need the knowledge base that will allow you, all of you in this learned body, to actually make wise decisions. We don't yet have that knowledge base. There are real issues with international governance, as you raised, and thank you. There are real issues with reception by people. I can tell you stories of—geoengineering cloud seeding that led ultimately to a professor from Penn State having a hole shot in his car door because the local farmers were very unhappy with the idea of cloud seeding. Sort of how this plays out into the broader populace is sometimes not as obvious and as simple as you might imagine. So I think gaining this knowledge base so that you would then have the capability of making wise decisions is wise.

Mr. MCNERNEY. Thank you. Again, Dr. Alley, the West Antarctic ice sheet has been noted to have the greatest amount of uncertainty in the melting and breaking rates. How much of the uncertainty related to West Antarctic ice sheet can be addressed by additional research, and how much is dependent on the future rates of warming?

Dr. ALLEY. Right. Certainly, the uncertainty can be reduced by the research, but it is already very clear that the faster and the more we warm, the more likely a failure will be. So in our world mitigation, trying to slow down the warming, buys you time. It buys you time to learn. There is always some danger with a tipping point that you pass it before you see it, and it's too late to slam on the brakes. It's too late to turn and avoid the iceberg. And very rapid warming, that becomes more likely for West Antarctica as we run at the future.

Mr. MCNERNEY. Well, what are some of the—Dr. Bell perhaps—what are some of the major concerns about the collapse of the West Antarctic ice sheet?

Dr. BELL. The major concerns are that it could go fast, and we don't actually know how fast. It's back to the ice sheet we know there used to be an ice sheet in New York and many of the States here. We didn't see that one collapse or the residents of New York then who didn't record what was happening. So we as a species don't have the record of how an ice sheet collapses, so we worry about how it collapsed—what happens to the ocean, how the ocean chews at the bottom of it as the ocean warms. We worry about what happens when the surface melt, where does that water go? Does it fall into cracks and act like a jackhammer to open it up, or does it run off like a river? There are some major fundamental understandings about how warming air, warming ocean impacts ice. And in that sentence alone you see how we have to have different disciplines talking to each other.

Dr. ALLEY. So Dr. Moon is working on this problem in Greenland, and Dr. Pfeffer is working on this problem in Alaska, as is Dr. Wolken, so the truth is the—what we learn spreads broadly.

Mr. MCNERNEY. Well, I hope the other three panelists don't feel neglected, but I only have 5 minutes, so I'll yield back. Thank you.

Chairwoman JOHNSON. Thank you very much. Mr. Babin.

Mr. BABIN. Yes, ma'am.

Chairwoman JOHNSON. Changing of the guard.

Mr. BABIN. All right, musical chairs. Sorry about that.

Dr. Wolken, in addition to serving here on the Science Committee, I also serve on Transportation and Infrastructure. And I represent southeast Texas and have four ports in my district. I recognize the importance of our navigational ship channels.

With that being said, one of the things I find very interesting on this topic that's relevant to my Committees and my district is the possibility of two trans-Arctic commercial shipping routes that are opening up. This isn't to say that I want to see all the glaciers melt and the sea levels rise uncontrollably, but if there are inevitable changes, I want to make sure that the United States is positioned to be economically fortified. And I know that the Russians are certainly exploiting newly opened up shipping lanes, ice-free zones, and even claiming certain areas that were considered in international waters are no longer that but belongs to Russia.

So how do you see the Department of Transportation or even the U.S. Coast Guard interacting with coordinated multi-agency collaboration that you say is needed?

Dr. WOLKEN. Yes, thanks for the question. I'll answer in two ways. The first is that what you speak of is really an incredibly important issue and, you know, economics and national security really do come to mind. And that's a sea-ice issue in the north, principally. And reduced sea ice of course is offering opportunities to enter into the Arctic and explore and ship, and that comes with fantastic opportunities of course and a lot of perilous conditions that could cause lots of environmental damage if not done right.

Having a multi-agency approach is incredibly important a little bit farther south. And you mentioned the Coast Guard. We have changes in Alaska that are impacting many of the fjords and the transportation routes in the South, and some of the changes in the cryosphere or changes in the snow, the ice, and the permafrost in the mountains are unpredictable to us right now. We don't have enough information. And so the Coast Guard communicating with various universities and agencies about how stable the slopes are, about how fast conditions are changing in certain areas could really be an asset to the Coast Guard as they respond to emergencies or possible disasters from cruise ships or fishing boats in different areas.

I will point out an example in 2015, there was one of the world's largest snow/rock avalanches into the Tyndall fjord, and in the process of that collapse, the tsunami that resulted from the rock falling into the fjord was enormous. It caused a trimline like the bathtub ring that was around 600 feet high. And any fishing boat caught in that or Coast Guard vessel or tourist ship would have been destroyed.

Mr. BABIN. Right.

Dr. WOLKEN. So communication about the data that we have to the individuals who will be working in these different areas, Federal agencies such as the Coast Guard, it's critical that we have this conversation.

Mr. BABIN. Absolutely. Thank you very much. And one other question. Some experts have predicted that our currently available mapping and navigation and ship capabilities are going to limit just how frequently and successfully we use these potential routes. And, Dr. Wolken, and to all of our witnesses, when conducting research on ice depth and volume, is there also efforts to improve commercial shipping potential such as data needed for mapping? Dr. Wolken, I'll ask you first, and then I'll go to Dr. Pfeffer.

Dr. WOLKEN. Yes, so a lot of the work that's being done in the fjords in Alaska are specifically focusing on the nearshore environment, and so the exchange of dynamics of interactions between the glaciers and the water in that environment. And so in the process of doing that, wonderful maps of the fjord are being generated; lots of different surveys of the coastlines are being generated in the process. And so the really great part about this is that we can have overlapping interests being served with good research in the right areas. And I think that's where this idea of having these interagency collaborations, these multiple perspectives, this team approach is really important.

Mr. BABIN. Great. And, Dr. Pfeffer, I think you wanted to say——

Dr. PFEFFER. Yes, I wanted to respond because your description of the situation in Texas reminds me a little bit of an experience that I had about 5 or 6 years ago where I was employed as a consultant for the Prince William Sound Citizens Advisory Committee, which is an environmental group that was set up in the State of Alaska following the Valdez oil spill to provide environmental oversight in Prince William Sound, which includes the town of Valdez and southern terminus of the Alaska Pipeline. And their specific concern was icebergs.

The Columbia glacier in Alaska, which is one of the glaciers that I've worked on for many years, was a major iceberg producer. And those icebergs came out into the shipping lanes. And the Alyeska, which is the operating company for the Pipeline, and the Coast Guard were both concerned about what future iceberg hazards were going to look like. Specifically, they had an ice detection radar system that had come to the end of its useful lifetime, and they had to replace it. And what their specific question was, you know, do we have to be worrying about icebergs for the next hundred years or the next 5 years?

And so I worked with them for about a 2-year period developing some simple models based on how much of the glacier was left and our best prediction of what the retreat would look like to give them some sense of what the iceberg discharge would look like. It was a good opportunity to collaborate with a State-level agency and also with the Coast Guard. We have a limited amount of bathymetry for that region. It would be good to have more, and NOAA has done some surveying in there. But that kind of interagency cooperation could be a lot more frequent than it is, and when it does happen, it's extremely beneficial. It certainly was a great help to us in Alaska.

Mr. BABIN. All right. Thank you. Thank you very much. I appreciate it.

Dr. BELL. I have a quick addition to that in that just last summer we saw one of the first groundings—I actually saw the vessel before it grounded of a Russian icebreaker that ended up grounding in the Northwest Passage, you know, exactly the places we're hoping or were thinking may be opportunities for more connections across the high Arctic. So it is a critical issue because it ran aground on an uncharted rock——

Mr. BABIN. Right.

Dr. BELL [continuing]. In essence. And the other piece is that the Coast Guard provides critical infrastructure to support the work we do in Antarctica. Without the U.S. Coast Guard and the heavy ice-breakers, we could not, the U.S. could not run the flagship programs they do. And we are seeing the Asian countries invest deeply in icebreakers. The Chinese Government has invested in two. The Koreans have a beautiful new icebreaker. We need strong ice-breaking capability both for ability to engage in the Arctic and continue to be leaders in Antarctica.

Mr. BABIN. And we have a shortage of icebreakers, do we not?

Dr. BELL. Yes.

Mr. BABIN. Yes.

Dr. BELL. That's why I thought this was a moment to remind you that——

Mr. BABIN. Yes. Thank you.

Dr. BELL [continuing]. With science, it's really clear—boats run aground, and we need icebreakers.

Mr. BABIN. All right. Thank you very much. My time has long expired. Thank you.

Chairwoman JOHNSON. Thank you very much. Mr. Casten.

Mr. CASTEN. Thank you, Madam Chair. Thank you all so much for being here.

Earlier this Congress at the Environment Subcommittee hearing on the impacts of climate change on our oceans and coasts, our experts were talking about what we need to do to stay below 1.5 degrees of warming. And I asked them if we got rid of all CO_2 emissions tomorrow, how much sea-level rise is already baked in? And the answer was an unequivocal 2 feet. I think that's consistent with your testimony, Dr. Alley. That is frightening, but in some ways, I have a bigger fear that's the deficiencies of our little Homo sapien brand.

And I want to demonstrate this and I want all of you out in the audience, you get to participate now. We're going to do a little experiment. So what we're going to do—this is real easy. I'm going to say two things. You give me the next in order. A, B? A little louder, come on. You got this. This isn't hard. Thank you. All right. Second one, 2, 4? You're all wrong. I was looking for 8.

This is the problem, right? We have all of these nonlinear trends, and our little brain says 2, 4, 6 and we see all these things that are going on. And, Dr. Alley, I think you alluded to this in your testimony. And so if 2 feet is baked in and if the likely skew of that data is not a bell curve but on the more frightening end of the spectrum, what sea-level rise should we be planning for within the zone of possibility?

Dr. ALLEY. I surely wish I knew. This is a frustration for us at a level that is deep and I wake up at 2 in the morning and I look at the ceiling and I say what do I tell somebody? I can remember coming back from Old Ironsides on the water taxi while doing the Freedom Trail in Boston and sitting in the water taxi and putting West Antarctica into Boston Harbor and not knowing what to do, which is—I mean, I'm sorry, it is very self-serving for me to sit here and tell you that funding research is good because it might go to me or my students, but we want to know.

Mr. CASTEN. So I'm not asking you for certainty, and I appreciate—look, I started my career doing—I got a master's of science in chemical engineering. I get the caution. But we've got to sit on this side of the dais and make decisions, so I'm just asking if you were in our seats with uncertainty of information, what is the range that we should be thinking about in our zone of possibility?

Dr. ALLEY. Yes. Don't go below the IPCC and start thinking about flexibility. Think about adaptive capability, the——

Mr. CASTEN. I'm just asking for like a number of feet.

Dr. ALLEY. Yes, I can't give you a number.

Mr. CASTEN. How about a——

Dr. ALLEY. I wish I could.

Mr. CASTEN. How about a timing? How long do we have before 2 feet is locked in?

Dr. ALLEY. Yes, very soon if not and, so for the 2 feet, you're getting close. But the big numbers, it really is, you know, I mean, a good businessperson looks for the black swan, but they don't know when a whole flock of black swans is coming, and so they really do look to their best people to be ready, which is here. You know, that's you.

Mr. CASTEN. I want to get to a couple other things in my time, but the reason I ask this question is in part because the same day that we had that hearing—I sit on Financial Services. We had Federal Reserve Chairman Powell in, and I said to him we just had this hearing. You are responsible for helping us write 30-year mortgages. Do you factor in whether or not those mortgages are going to be paid off in low-lying coastal areas? And the answer was that he thought we probably should start thinking about that but we haven't yet.

We have a whole host of issues here that go just beyond whether the sea level's a little higher, right? We got housing. I live in Illinois where we've got, you know, polar vortex because—and polar bombs or whatever the term is of this year because, as that ice melts, we're destabilizing global weather flows and shifting that cold air down temporarily until we all get a lot hotter.

Dr. Wolken, I had a little fun doing a little Googling on the weather report on Moose Mountain where you live. I understand you got a huge unseasonable amount of rain a few hours ago. I understand that is pretty positive because you've got some concerns up there. Can you just help explain to me what's happening on Moose Mountain that makes that rain good and how that is related to the falloff in sea ice?

Dr. WOLKEN. Wow, that's a really good question. I will preface this with some history about the winter. It was a very low snow year. We didn't have near the snow that we would normally have. And this is a trend especially across the Arctic. And this year it's been unseasonably hot. In fact, this week in Alaska many records have been broken. And this is common as well in recent years.

I left Alaska the other day to evacuation notices, so before I came here, we were planning to evacuate our house because fires were raging just near our house. And so the rain coming is a great idea. The whole State is suffering from smoke right now because there are so many fires really resulting from a chronic low-snow issue and having warmer temperatures that are really fueling the fires. And so this is a major issue for us, and it's become quite personal for me.

Mr. CASTEN. So just last question for the whole panel, has anybody estimated how many people's homes are at risk because of this combination of sea-level rise, spreading wildfires, flooding in the Northwest? How many people do we need to be thinking about dealing with right now? Do we have any estimates of that answer, Dr. Moon?

Dr. MOON. So I'm going to give you an estimate that's just a fraction of those things that you just asked about. This is just an estimate on homes. It doesn't include power plants, airports, military bases, anything else, just homes. If we're looking at 1 foot by 2035, that would be about 140,000 homes. If we're looking at 4 feet of sea-level rise, that's about 1.2 million homes. If we're looking at 2 feet, that's about 300,000 homes. So it's in the hundreds of thousands, and if we look at levels where we're reaching 6.5 feet of sea-level rise by 2100, we're looking into the trillion-dollar kind of mark just for homes. That's not other roads, other infrastructure, et cetera.

Mr. CASTEN. And I would presume that's just coastal. That doesn't include Dr. Wolken's house that may be at risk——

Dr. MOON. And it doesn't include wildfires or any of those other things that you mentioned that will be also addressed by addressing climate change.

Mr. CASTEN. Thank you. I yield back.

Chairwoman JOHNSON. Thank you very much. Mr. Baird.

Mr. BAIRD. Thank you, Madam Chair and Ranking Member. I guess my question deals across the board. We got all doctors here as witnesses, so the question I have simply, what Federal programs are most critical to gaining a greater certainty on the future change in ice sheets and those effects on sea level? So you can go in any order you want. Dr. Alley, you want to start?

Dr. ALLEY. Right. It is interagency. The National Science Foundation provides so many of the people support, and they do the lead agency in Antarctica and in some other things. NOAA, we have to have what they are doing. NASA has been keynote not only on Operation IceBridge, which we have been talking about, but the satellite monitoring. The DOE has a role in modeling, and so I've hit a lot of the high ones, but it really is the interagency, the U.S. Geological Survey. When I gave the number on how rapidly the icebergs were breaking off when John Muir was watching, that number came from the United States Geological Survey. So it is having these wonderful centers of excellence that you have built that live in the U.S. Government and give us leadership, they are not localized in one place. They are in several agencies, and they work together, they know each other, and they can do this with support.

Mr. BAIRD. Thank you. I'll remind you that Madam Chairwoman gave me 5 minutes, so we can spread that out.

Dr. BELL. Well, I will echo the NSF for understanding why, NASA for monitoring how it's changing, the USGS for incredibly important measurements of the glaciers, DOE for modeling, and NOAA for lots of information about how the ocean is changing and what the fundamental tide gauges are doing.

Mr. BAIRD. Thank you.

Dr. MOON. You asked about narrowing our range of what's going to happen into the future. On the science side, integrating better observations and understanding of the physical system into our models, our models can't make up that information on their own. But I also want to reiterate that it is our mission's pathway that is going to make a tremendous difference in what that future number of sea-level rise and our future number for ice loss is. That's not the science part.

Dr. WOLKEN. Yes, I just want to echo the comments of my colleagues here and really just add that we're doing this in Alaska already. We're getting as many people together as often as possible to try to solve some of these issues. The only way to really do this is through an interagency perspective. And there's really no other way to address such a large issue. And all of the Federal Government programs are critical to what we do.

Dr. PFEFFER. OK. Well, again, I'm echoing what all of my colleagues have said, but I want to add to this. The problem of collaboration and communication between these agencies is not an easy task. One example, NSF operates on a principle that could be summarized as turn the brightest people loose on the most interesting questions. The fundamental function of NSF is to support these investigator-based science where each one is evaluated on its own scientific merits. It's not a mission-driven agency in the way that, say, NASA is. That has produced—it's been extraordinarily successful by letting scientists decide on what's the best thing for them to study.

But in a situation like sea-level rise, I think that more—well, it's for climate change generally, not just sea-level rise, I think that a more coordinated approach is necessary. Back in the early 1970s the National Science Foundation had a brief program called RANN, Research Applied to National Needs, where basically a management structure was experimentally imposed on research programs. And it was a notable failure. Almost everybody that you talk to that knows about RANN say, oh, boy, yes, that was a bad time at NSF but not everybody.

It's a little bit like the Manhattan Project. If the Manhattan Project had started out with the, you know, advisor saying, OK, we need to understand about atomic energy, all of you pick an interesting problem and go work on it and come back in 5 years, you know, that's not the way the Manhattan Project worked. And I don't think we're going to solve this problem that way either. I'm not talking about the magnitude of the project or how much money should go into it, but I am talking about coordination and the need for some really innovative thinking about how those agencies should interact because it's hard to steer scientists and, you know, it really is a herding-cats problem.

But particularly with all these agencies, there needs to be some really imaginative way of figuring out what gets done first and how long do we have to solve it. And I don't have any answers to that, but I think that's a really strong need.

Mr. BAIRD. Thank you, and I'm out of time and I yield back.

Chairwoman JOHNSON. Thank you very much. Ms. Wexton.

Ms. WEXTON. Thank you, Madam Chairwoman, for yielding. And I thank the witnesses for appearing today.

It has been quite alarming to say the least to read your testimonies and also listen to the responses to some of my colleagues' questions here today. It is absolutely clear to me and it should be clear to everyone that we are at a tipping point of our Earth changing dramatically and irreversibly due to human-caused climate change. It's even more alarming that we're locked into 2-feet of sea-level rise—everybody seems to agree about that—and that, given the melting of the Greenland and Antarctic ice sheets, we could be looking at 11-feet of sea-level rise.

And how you prepare for that is something that is really important to all of us and certainly to me in my home State of Virginia. We have a lot to contend with over that. We are home to Naval Station Norfolk, which is the largest naval base in the world, and Langley Air Force Base. They are already having to deal with the effects of sea-level rise and the effects it has on our national security. And I'm also the mom of two kids, and I worry about what kind of a planet we're leaving for them and their kids.

I know that we had some questions about climate ice cover and sea levels and how they routinely change from season to season and over time. Some claim that this natural variability means we shouldn't be concerned with humans changing the climate.

Dr. Alley, I know that Mr. Brooks asked a little bit about this and Dr. Pfeffer did give some explanation of what is actually happening. But, Dr. Alley, can you explain what the science tells us and why we should be concerned with the changes in ice and sea level and climate that we're seeing right now? What makes it different from what happened, you know, over the past 20,000 years?

Dr. ALLEY. Yes. So thank you very much. It's wonderful that people take interest in what we do, you know? So, as you know, on a dry, hot summer day, you know, the hills of Virginia have always burned when there was a lightning storm. And because you know that, if you see kids headed out on a dry day with illegal fireworks, you were very worried about it. We know that people have always died, so we have metal detectors at the front of your building here. We know that climate has always changed, and that proves that climate is changeable. And you've never met the person who said the hills have always burned, so we won't worry about arson. But you have met the person who said the climate has always changed, so we won't worry about humans changing the climate.

The climate has always changed proves that the climate is changeable. The climate change has always affected living things, which proves it's important. Climate has changed for a lot of reasons, but CO_2 has been especially important. And that points a finger at us. Now, if you were an arson investigator, you better know natural fires. You do CSI (crime scene investigation) fire. If you're a homicide investigator, you do CSI homicide. We do CSI ice. We do CSI climate. And we actually have very high confidence that what is going on now is human, not natural. If anything, over the last small number of decades nature has tried to cool it off a little, so how much of the warming has been us is a little bit more than all of it is the central estimate.

But the fact that nature has done these huge things in the past, that when nature warmed a little bit, sea-level rose a lot. And then you say, well, we could cause a whole Ice Age of warming with our CO2 in the future. And the last end of an Ice Age gave us 400-feet of sea level. There's 200 more left.

So I believe that climate has always changed is a very, very strong argument to be concerned about what we're doing for climate in the same way that burnable hills make you nervous about arson. And when——

Ms. WEXTON. And related to that, Dr. Alley, in your testimony you discussed several studies that suggest that the IPCC report is overly conservative and underestimates the rate at which ice sheets are and will continue to melt.

Dr. ALLEY. I have great difficulty finding any evidence that they are overly alarmist, and there certainly are things that point to the possibility that they have been low in the past. And, yes, that's fairly clear. When you look at the history of——

Ms. WEXTON. Can you discuss this current scientific research on estimates for tipping points for the Greenland ice sheet, Arctic ice, and

Antarctica ice? What are the tipping points or what does the science tell us?

Dr. ALLEY. So Greenland, as it gets thinner, it gets warmer. As it gets warmer, it melts faster and gets thinner, and at some point it will be committed to loss. It probably will melt fairly slowly. West Antarctica, if it starts doing what the glaciers in Alaska have done, the coastal glaciers have done, it could go very, very rapidly. We're cautiously optimistic that the sea ice in the Arctic will act like a dial rather than a switch, but we're not entirely sure of that. We are worried a little bit about circulation in the Atlantic and other places that act more like a switch or a tipping point. The National Academy of Sciences looked at tipping points in 2013. They especially pointed to tipping points in ecosystems and in human systems. So at what point when people are stressed and they're having to move their houses or change what they do, at what point do the people become very mad and then tip into some other level of behavior. And so when you look, there are some physical tipping points, there are more ecological tipping points, and there may be a whole lot of people tipping points.

Ms. WEXTON. Thank you very much. I see my time——

Dr. ALLEY. Thank you.

Ms. WEXTON [continuing]. Is expired. I yield back.

Mrs. FLETCHER [presiding]. Thank you. I'll now recognize Mr. Gonzalez for 5 minutes.

Mr. GONZALEZ. Thank you, Madam Chair, and thank you, everybody, for being here for this important hearing.

I want to focus at least the beginning of my time on adaptation and resiliency. I think it unfortunately seems like there's a lot that's sort of locked in that we're going to be dealing with over the next however many years. And I'll start with Dr. Pfeffer. What are you seeing or what guidance can you give us with respect to making sure that we can adapt as sea levels rise and that we're building more resilient infrastructure?

Dr. PFEFFER. So I mentioned in my early comments I really am concerned with—in the work that I've done in the near term, the next 30, 40, 50 years where this constellation of factors has to be considered. One of the very interesting and extended conversations that I had was with a man named David Behar, who works for the San Francisco—the city of San Francisco as a coastal engineer. And one of the problems that they have to deal with are—it's a very large dike system that basically surrounds San Francisco Bay. And they need to know how far do they have to raise this dike system, which is extremely expensive? It's in the billions of dollars for a very small rise. And so it was not adequate to simply say, well, let's just be safe and figure on 10-feet of sea-level rise and then, you know—and you only get 1 foot and you've spent an awful lot of money.

Mr. GONZALEZ. Yes.

Dr. PFEFFER. In the same sense, one of the questions—and this goes back to an earlier question about how many people may be displaced by sea-level rise. If you take an overly conservative number meaning let's take worse-case scenario and you draw a line on the coast saying, OK, this is going to be inundated by such-and-such a date, what happens to the value of those homes on the basis of that line that you've drawn? And the nearer in time you get, the more important that becomes. So you really have to have a tight bound on sea-level rise and a tighter bound to the nearer to the present that you get. We don't really have that yet. In some places we do, and it's—very often is a group of scientists that live in a particular region like Hudson River, for example.

Mr. GONZALEZ. Yes.

Dr. PFEFFER. San Francisco Bay is another example where you can look at all of the causes of sea-level rise, including things like isostatic depression or rebound in an area as—partly as a result of large-scale things like ice sheets disappearing 20,000 years ago and partly local things like putting buildings on that land.

Mr. GONZALEZ. Yes.

Dr. PFEFFER. There are a lot of different factors that have to be considered and different time scales you deal with different factors. And I think it's another thing that points to this interagency collaboration.

Mr. GONZALEZ. Got it.

Dr. PFEFFER. But one of the things that I've tried to emphasize in the past is there's certainly a cost to neglecting sea-level rise, but there's also a cost to overestimating.

Mr. GONZALEZ. Yes. And I think that's actually a really important point is, you know, when we talk about resiliency and adaptation, there is a cost to all of this, right?

Dr. PFEFFER. Yes.

Mr. GONZALEZ. And we can't completely ignore that. We can't be too conservative or too aggressive or——

Dr. PFEFFER. That's right.

Mr. GONZALEZ [continuing]. You know, we're going to be wasting a lot of money.

Dr. Wolken, if I could shift to you quickly, in your testimony you mentioned that in Alaska there are only three long-term continuous records of glacier mass for the entire State. Considering remote sensing and computer modeling are used to predict future scenarios due to the lack of ground-based observational data, how reliable and accurate are remote sensors and computer modeling in measuring glacial melt and predicting future changes?

Dr. WOLKEN. Yes, we're doing really well with these different tools, and I think, you know, one of the things that you can envision is if you go to the hardware store and you get a laser range-finder, for instance, from the shelf and, you know, you do some home renovations at your house, well, that laser is actually quite accurate. It's a laser, and it's very precise and accurate. And we use tools like that to really gauge how the ice is responding. We use other remote-sensing tools to do similar things, to see how much it's changing in this direction. And those are incredibly useful, and that's how we do things. We do those with both airborne and satellite-based assets.

There is a need in places like Alaska where the topography is so extreme and where the changes are so great to actually have ground observations. And so when you're using these different remote-sensing tools, the resolution isn't quite there some of the times, and so having ground observations to validate in some way or to correct in other ways is really the way to go. And so more ground observations truly do help us. With a lack of that, we have no option but to use the tools that are in front of us, and really, remote-sensing-based opportunities are where it's at for us.

Mr. GONZALEZ. Great. Thank you, and I yield back.

Mrs. FLETCHER. Thank you. I'll now recognize Dr. Foster for 5 minutes.

Mr. FOSTER. Thank you, Madam Chair. And I'd like to thank really the Ranking Member and all Members of this Committee and the witnesses about the tremendous increase in the level of serious discussion that we're having on issues like this over the last 2 years.

I think if you Google my name along with Greenland, you're led to a video of a previous witness who was a lawyer trying to convince this Committee that it was a matter of scientific debate whether or not it was a good thing that the Greenland ice sheet melted, OK? And so we're having a long-overdue and very high- quality discussion here.

Now, my next question, how many of you knew Charlie Bentley? Wow.

Dr. ALLEY. He was my Ph.D. advisor.

Mr. FOSTER. Well, oh, wow. I grew up next door to Charlie Bentley on Lake Mendota in Madison, and, you know, and I remember sitting on his porch discussing what he did. You know, he would disappear every couple of years and study the ice sheet in Antarctica, which did seem goofy. And I think it's a lesson on curiosity-driven research, that this thing, over the course of his career, went from something that was done by, you know, sort of an eccentric professor to something that is now going to be an absolutely crucial thing in deciding how we deploy trillions of dollars of capital to try to mitigate the damage of this.

Charlie passed away I think a couple years ago, and I understand there is a mountain named for him in Antarctica. Anyway, I was pleased to see the recognition among the Committee here.

Now my next question I had is, what is known about the speed of response of the ice sheet system to changes in temperature? You know, there are natural experiments when you get volcanoes going off with a couple degrees swing for a few years, is that long enough to actually be seen in the response of the ice sheet?

And the reason I'm asking this question is I think it's likely that we'll be able to decarbonize the U.S. economy. I think it is much less likely, you know, since we're 5 percent of the world population that we're going to be able to convince the rest of the world to decarbonize as quickly as necessary. And if that happens, then I think it's likely we'll be looking at things like albedo modification which has the potential of very rapidly changing the temperature. There's an article in Nature earlier this year that used state-of-the- art climate models to say, OK, you know, will it work or are we going to get cyclones and so on? And the first look was that it might be feasible.

But they didn't, to my remembrance, model anything having to do with the ice sheet. And so I was worried that maybe there was sea-level rise locked in just due to the thermal time constants, that even if you rapidly bring down the temperature of the atmosphere, that it will take a while. And so what is known in modeling or in data about that issue?

Dr. BELL. The ice sheets respond slowly—they're slow. I mean, when Richard and I started studying ice, we couldn't imagine they'd change as fast as they are today. I mean, Charlie actually— one of my first papers I wrote told me I couldn't write that they were going to change fast because even in the 1980s we couldn't imagine the speed at which we're seeing now. And now you can actually occasionally hear fear in scientists' voice because they are changing faster than Charlie thought they could when you grew up next to him, that we just couldn't imagine these thick pieces of ice changing, and he couldn't either.

But now we know they're changing due to the ocean warming and the atmosphere. The atmosphere is a faster driver than the ocean. So it will—there's—we don't have a good handle on how fast it's going to respond—

Mr. FOSTER. Do you have models that even make a decent approximation? Can you see, for example, in response to volcanic eruptions and the swing there, can you see changes in the rate of ice accumulation or de-accumulation?

Dr. ALLEY. We can do a lot of it right and a lot of it not yet, and so I could brag on the progress that we've made and some of it with Charlie's help. And I could bore you or scare you with where we're missing, especially the couplings into the ocean. So if you start blocking the sun, what it does to the atmosphere is fairly straight- forward. What that does to the ocean, which is interacting with the ice, is not at all straightforward, and that really needs work. And there is——

Mr. FOSTER. Are these computing-limited problems or knowledge-limited problems?

Dr. ALLEY. Yes, especially knowledge-limited. The computing is coming. We could use a little more, but it's primarily knowledge- limited.

Mr. FOSTER. Yes.

Dr. BELL. It also has to do with those measurements, our lack of knowledge of even what the ocean temperature is around Antarctica. We can look from space—we can measure the top of the ocean, but we're still so limited in understanding what's going on at depth, and that's what matters because the critical parts of the ice sheet that are really—the sensitive switches, those are down low, and we don't have the good measurements.

Mr. FOSTER. Let's see. No, I'll abuse another 20 seconds. Yes, for the last——

Dr. PFEFFER. Yes, I wanted to add, and it's already been mentioned that the IPCC's fifth assessment, their discussion of sea-level rise is very conservative. I was one of the lead authors on chapter 13, which is the sea-level chapter, and that discussion that we had about what number are we going to put in for our upper limit, and I remember that very vividly. And essentially what we did is we said we just do not know yet enough about the rapid tipping point mechanics to be able to attach a number to this rapid response.

Mr. FOSTER. Thank you. And to the extent these are compute-limited modeling problems, you're very welcome to use the super-computers at Argonne National Lab in my district—I can't think of a better use for them.

Dr. PFEFFER. Yes.

Mr. FOSTER. Thank you. Yield back.

Mrs. FLETCHER. Thank you. I'll now recognize Ms. Stevens for 5 minutes.

Ms. STEVENS. Thank you, Madam Chair.

We got to talk about the psychology here. I just went from another hearing on heat effects in the workplace and the overheating in the workplace from warehouses to fields and how that's impacting human health. We're willing the have the dialog on climate change.

So, Dr. Bell, you had a chart that kind of showed the sea levels and made that point about your father's life. How long have we been able to actually talk about rising sea levels and their impacts on us? How has modern science been able to influence this discussion and the question of what we can actually do to combat this?

Dr. BELL. The answer is, you know, people have been living with changing ice for a while, but the real understanding of the linkage between the changing ice—because people who live up near the mountains like the people who live in Alaska are very aware of the changing ice and the people who live at the beach. It's really I'd say in the last two decades that it's gotten very strong, that connection. In fact, back to your psychology point, it's really in the last decade that we're starting to see the conversation about the psychology of how we handle it.

And it was only this year that—actually last month that the Earth Institute handled the—convened the first-ever conference on could we ever talk about managed retreat? What would that mean? How do we have that conversation? So in fact we are just opening this door of connecting the work that we all do on frozen stuff and beautiful places far away with what happens to our assets at the coast, to our beaches, to the naval ports, to the airports. We happen to like to build airports. You know, of the 10 impacted airports on the planet, five of them are ours because that's a good place. So we're just starting——

Ms. STEVENS. Well, and we're coming up with new terms, I mean, with these extreme weather events and what it means. And, you know, we can study the free-rider principle. We can study, you know, you start to think of like nuclear warfare or weapons of mass destruction and when faced with that threat and what do you as an individual do. What do you as a society do? What do we as a body of Congress do? What is it going to take for us to take this seriously?

I'm in Michigan, and I don't have a lot of sea around me but I got a lot of lake. And this is going to impact us. You know, I stumbled across a video. What is a world without ice? Is it going to take a modern society to see a full city go underwater for us to take climate change seriously, for us to take rising sea levels seriously and the grand challenge that we actually can do something about it, that it's a uniquely positioned challenge for us as the great America to take on?

So I don't know who else, with all your great expertise and your phenomenal science and all your great background can provide some guidance here, some common sense for us to not just talk about it but to do. And I don't know who can chime in here because we do this on recycling, with the plastics crisis, and what the individual can do, what the body can do, the body that we're currently in, and then on. Thank you.

Dr. ALLEY. Right. So you raised very important questions. I wish we had good answers. But you know the Nobel Prize in economics, corecipient last year, William Nordhaus from Yale developed tools which allow decisionmaking or inform decisionmaking. And he showed that efficient response on climate change helps the economy, right? If you want a bigger economy with more jobs, you take actions that honor the science on this. Many of our medical professionals, through their organizations, have said this is a serious health issue, that actions that would reduce the warming will have health benefits. Our military leaders have been very clear on the national security issues of not dealing with this. So environment and ethics are actually in the direction of economy and employment, as well as national security and health. And we can see futures in which very expensive sea-level rise happens and large other changes happen, and we can see very bright futures where we use our knowledge.

The Yale climate communications people have surveyed what America thinks about climate and climate science. Most Americans tend to accept climate science but not all. But many, many, many Americans are very excited by the solution space. And if you ask them should we solve it, even if maybe they're not sure there's a problem, they're happy to go look for solutions.

Ms. STEVENS. Yes.

Dr. BELL. Representative Crist, I used to always show pictures of drowning Florida in my presentations. You know, I'd show how much I'd drown Florida. I drown Florida a lot. I decided I couldn't do that because it was depressing people, turning them off, and they were not listening. I see change—we have to move—before we drown Florida or New York where I'm from, we have to actually start thinking about how we as individuals—I very much worry about what I as an individual—I worry about my community I live in. I worry about my professional society, what we can do. And I'm very happy that you are asking—I worry about our local government, and I'm very happy that you are engaged—and it's essential that you take leadership on this, too, because we can lead if we step forward before we drown a city or a State.

Mrs. FLETCHER. Thank you. I'll now recognize Mr. Crist for 5 minutes.

Mr. CRIST. Thank you, Madam Chair. I appreciate the opportunity. I have some very well-prepared questions that my staff has put together for me, but you've inspired me to kind of go off script. I am from Florida, and my colleague asked I thought an excellent question about, you know, what is it going to take before, you know, Congress takes concrete action, each of us as individuals do so.

I live in St. Petersburg, Florida—it's on the west coast—and have lived there since 1960. And I live downtown. My parents live in the northeastern part of the city. We both live on the coast. So when I drive to visit my parents, who, thank God, are still alive, I go along the coast and I can see the difference in the sea level in the bayou that they live on. And I have noticed it significantly greater in the past 5 years. And I don't think that's just an anecdotal thing. I think it's a real thing.

You know, I previously served as Governor of my State, so I would travel the whole State quite a bit. And whenever I would go down to Miami, on Miami Beach in particular, there's a road called Alton Road. And Alton Road will flood when it's not raining. And I remember President Obama visiting south Florida and would talk about that example of, you know, the climate changing, the rising sea level. And so we

Floridians get it because we've seen people drown a lot. And we're witnessing that occurring, you know, so it kind of freaks us out.

And, you know, it seems to me that we need to figure out a way to sort of get off the dime. And I'm sure, given your illustrious professions and dedication to what you study that it's got to be frustrating for you as academics to not see a whole lot of action in this area.

And I'm going to ask almost the same question but maybe in a different way. What kind of advice can you give us—as hopefully decent communicators to Americans—to motivate action?

Dr. MOON. So I want to reiterate Dr. Alley's point in emphasizing solutions. I had an opportunity to give a TEDx talk to my community in January, and I emphasized the solution space of it. And I had a friend then a month or two later sent me an article about the U.N. report, which told you about all the horrible things that are coming down the Pipeline for us. And she said is this true? This seems really radical. And I said, yes, all the information in there is true. And she said, you know, this hasn't motivated me at all, but your talk did.

So I want to emphasize that we need to be talking about solutions. That's motivation to people who don't even necessarily think about climate change because they wanted to be getting renewable energy, becoming energy independent, which is something that we can do with that sort of thing. The solution space is very inspiring. As Americans, we have led, we have innovated, we have created new paths for the world, and I think that we can convince people that we can do that in this space as well because, in fact, we can do that in this space.

And then the one other thing that I want to say in this area, too, is that it's about encouraging people to talk about this and come together with each other, too. We simply don't talk about this enough. And if we talk about solutions, we can also think about how we're directly helping people. I mean, in the last couple weeks we've heard about hundreds of people being laid off from coal mining jobs because of bankruptcies or other problems the—in decline in coal. But if we're thinking aggressively about moving forward, we can think about how are we going to give these people other jobs? How are we going to support them as we're losing this industry instead of just putting our head in the sand as we lose this industry, which is hurting people on both sides.

Mr. CRIST. Well, if I could follow up, I have a little time left. In speaking about the solutions, what are the most obvious ones to you that you would be willing to share with us?

Dr. MOON. Well, I'll tell you, I'm a scientist, so in my personal——

Mr. CRIST. Thank God.

Dr. MOON. In my personal solution space, a lot of it is in communication. I don't envy you as policymakers and having the much more difficult job in discussing all of the elements, not just science, that go into your policy decisions. And unfortunately, many of the questions on those solutions lie on your desks. And I really would love to see us depoliticizing climate change so that all of you can spend your time discussing which of these solutions we're going to implement and how.

Mr. CRIST. Thank you very much. I yield back. Thank you, Madam Chair.

Mrs. FLETCHER. Thank you, Mr. Crist. I'm going to recognize myself for 5 minutes, and then we'll continue with the hearing. This has been a really great panel, so I want to thank all of you for the time that you've taken with us this morning. And I want to thank Ranking Member Lucas for holding the hearing. We've heard great questions and great answers.

And there's a lot that I'd love to cover, but, Dr. Moon, I want to go back to something that you talked about in your opening. You were talking about sea-level rise and sort of potential possible rise levels. And you mentioned that it was going to be 2.5- to 3-feet potentially in the next 80 years, and that number would be higher on the Gulf Coast. As a Representative of Texas' 7th congressional District in Houston right off the Gulf Coast, that of course perked up my ears. And I would love to learn more about why this is, what is the best estimate for the Gulf Coast region? Certainly, all of our eyes are on New Orleans right now. All of us are focused on the impacts of hurricanes and overall sea-level rise on our coastline and the Gulf Coast. And so part of the question is why is it and also what can we do about it?

Dr. MOON. There are a variety of things that determine sea-level rise in your local spot. So where we're losing ice on the Earth makes a difference. You are going to be influenced differently by losing ice in Antarctica than Greenland. There's also the ways that ocean currents and atmosphere currents move, pushes oceans one way or another, and also what's happening in your local region as far as your land naturally rising or falling already. And that's a— land subsidence is something that we see broadly across the Gulf Coast.

So there are these multiple different elements that all stack up to make what you in your individual city are going to see as far as sea-level rise. And it's quite consistent that in the Gulf Coast region we will be seeing substantially more than the global average over the—since roughly 1960, many areas along the Gulf Coast have already seen 8 inches or more, which is much more than the global average during those periods.

Mrs. FLETCHER. Thank you. That's helpful, helpful information. And something that we do talk about a lot and we talk about resilience and rebuilding a resilient infrastructure, there are a lot of issues, and I think it is top of mind in a way that it might not be for some other folks in terms of sea-level rise. But I think one consistent theme I've heard from every witness today is that we need more people doing the research, helping us get the information that we want to know so that we can make smart policy decisions and that we can know what we're dealing with.

So I really want to put this out to the entire panel to talk about how we are recruiting and training the next generation of glaciologists and where there's room for us to help. What kind of policy can we implement here, what kinds of things can we do in addition to funding that would be helpful for you, and for anyone who wants to take that on and talk about what we can do to increase that number from 1,400 to—and maybe what number you think would be good overall.

Dr. BELL. Well, there are 13,000 people who are members AMS just to give you an idea of what—who—and that's the American Meteorological Society, so who's working and worrying about the weather in the U.S. We have 13,000 people doing that, and we have 1,400 around the globe doing ice. So numbers should be higher than 1,400. Let me give you an order of magnitude.

What can we do? I think it's partially making it so everybody can talk about the science and back to Twila's point about it not being politicized and also making it so—I think we're driving some of the young talent from the field because it seems like it's a hard place to be, not because it's hard to go to the field and see beautiful places but because it's hard because you're under attack.

I think embracing science so we have within our communities science-based, evidence-based planning for the future I think will attract more people because young people want to make a dif- ference in the world. And if they see there's science, even if you're studying how ice deforms and flows, is going to matter to what happens in your district, that's one way we can help attract it, by working on—even by holding this hearing is huge, but by working to ensure we have scientists intimately involved with developing the policy on how we're going to lead in the future.

Mrs. FLETCHER. Thank you. Would anyone else like to weigh in on that?

Dr. PFEFFER. Yes, if I could add——

Mrs. FLETCHER. Dr. Pfeffer.

Dr. PFEFFER [continuing]. A couple of comments to this. I mean, I think we're suffering to a certain extent from sort of, you know, the boiling frog syndrome of things changing around us at the moment at a rate which is, you know, gradual enough that we can say, oh, you know, this is just sort of natural variability or I remember something like this happening 20 years ago. I think making climate change generally a reality for people involved, somehow bringing it out of sphere of scientists. You know, a news report will say, OK, here's a scientist in Antarctica who has done such-and-such and thinks this, and then they show the picture of an icebreak or something, which to the ordinary, you know, person on the street, it looks like these scientists are on a different planet. It's all kind of removed from them and—in the hypothetical. And somehow this link—and I think things like this hearing are creating this link that's not just scientists in this hypothetical space discovering this thing which can only be detected through sophisticated measurements but that it's actually happening in a way that everybody is feeling, and it's happening now. We're no longer waiting for the evidence that climate is changing. We've got it. We've got buckets of it. And that boils down to communication. And I've done a lot of public presentations. As was mentioned earlier, I was involved in the movie "Chasing Ice" and have done a lot of that kind of public communication both before and after, and very often I get questions from people about, you know, what can we do? And it can be very hard to answer that question, especially if they're asking what can I do personally about climate change because it just seems like such a big problem? And one of the things that I do say to them is, you know, things like installing fluorescent light bulbs and, you know, buying a more fuel-efficient car doesn't seem like much, but we did create the problem one airplane seat at a time, one car at a time, one truck at a time. And the individual action does matter if everybody does it. And so recruiting people to understand and accept that this is a reality is sort of the first step.

Mrs. FLETCHER. Thank you. I yield back my time, and I'm going to recognize Mr. Tonko for 5 minutes, who will then close the hearing. Thank you all very much for your time today.

Mr. TONKO [presiding]. Thank you, Chair Fletcher, for what I think is a very important hearing. Thank you to the panelists for setting such a respectful tone for science, refreshing. I represent New York's 20th congressional District, upstate New York, and it's home to much innovative pioneering work, the topic before us. At Union College in Schenectady, for example, Professor Rodbell has been working for more than 30 years to document glacier fluctuations in the Peruvian and Ecuadorian Andes. Professor Rodbell and his students are conducting ongoing research on glaciation in the Andes with a specific focus on determining rates of current ice retreat compared to natural rates of ice retreat in the geologic past. At the University of Albany, Dr. Mathias Vuille, a professor in the Department of Atmospheric and Environmental Sciences, is researching climate impacts and glacier retreat in the tropical Andes. In February of this year, Dr. Vuille testified in a public hearing held by the New York State's Senate Standing Committee on Environmental

Conservation. He noted that sea-level rise is resulting from warming of the ocean and added water mass due to ice melting glaciers and ice sheets in Greenland and West Antarctica. He noted in particular that sea-level rise is not equal everywhere and sea-level rise in the mid-Atlantic and New England coasts are much larger than the global average. He also emphasized that since we have no glaciers in New York State, impacts can seem far away and irrelevant, but glacial melt affects us nonetheless.

So, Dr. Bell, can you describe the indirect impacts of glacial and ice sheet melt in States like New York that I represent that do not have glaciers?

Dr. BELL. Well, thank you very much for that question. I'm also from New York, so—and the ones I'm going to speak of are—actually the nice examples are in New York because—because of sea- level rise, the number of people impacted by Sandy was significantly larger because—because of that in New York it's about 9 inches in the last hundred years the sea-level has—you can see the record right from the Battery. And you can see how many more homes were flooded, how many more people were impacted, and today, we're seeing that those are the homes that are actually being built up along the edge of the Hudson. It now looks like you're in New Orleans. The homes are being elevated right there in Haverstraw. You have homes that you could see in New Orleans. So that's the kind of impacts we're seeing. You're seeing that we've had Sandy. We impacted far more people, tens of thousands more people than we would have, and now we're responding to it.

Mr. TONKO. Thank you for that. And what more can you tell us about the uneven distribution of sea-level rise across our country? What will sea-level rise look like, for example, on the East Coast versus the West Coast or in New York City versus Washington, D.C.? What are the wide-ranging impacts of sea-level rise?

Dr. BELL. The National Climate Assessment did a beautiful job of laying out those variations and going through the different parts of the U.S. and really explaining the difference. But briefly, each community has to worry about which ice sheet you're close to. If you're close to an ice sheet, it turns out it doesn't matter as much, so for New England, Greenland—

Antarctica matters way more than Greenland, whereas the Representative from Florida is going to see both Greenland and Antarctica full on. So there's the proximity. Then there's how close you are to ocean currents. That's some of the change we've seen in New England is the warming ocean has impacted New England. And then the Representative from Virginia is seeing the tremendous impacts of local subsidence around Norfolk because you've withdrawn water, so the land is going down at 4 millimeters a year. You add that onto the sea level going up, suddenly, you have a problem.

Mr. TONKO. And to anyone on the panel, what mountainous regions around the world are most at risk, and what adaptation measures can be taken to avoid large flows of environmental refugees?

Dr. PFEFFER. If I could——

Mr. TONKO. Yes, Dr. Pfeffer.

Dr. PFEFFER [continuing]. Address that, there are potential for environmental refugees at sort of both ends of the hydrologic cycle. Let's discuss the Himalayas, for example. Earlier, I mentioned the various geologic hazards that people in the immediate vicinity of glaciers, these high valleys, high density of people in those valleys. As we go downstream, there are people who are very dependent upon runoff from those mountains for crop irrigation, so this goes out of Nepal and into India. And then the people on the coast—and Bangladesh is very often used as the example—that are at risk from sea-level rise.

So everything from geologic hazards to changes in water supply to sea-level rise, each one of those has a population which is put at risk. And as far as mountainous regions where this really matters, certainly the Himalayas, also portions of South America, Alaska is subject to certain risks, but the primary influences there I think are going to be environmental on the changes in water and immediately coastal effects.

But the people I think really in the Indian subcontinent, they're at very high risk, and that is a global problem. It's not just a problem for them, and I think that's probably very clear.

Mr. TONKO. Thank you. Thank you very much.

Look, that concludes, I believe, all who have chosen to ask the witnesses any questions. Before I bring the hearing to a close, I do want to thank this panel. Thank you so much as witnesses for testifying here before the Committee. And I want to thank both our Chair and our Ranking Member for hosting this hearing. The record will remain open for 2 weeks for additional statements from the Members and for any additional questions the Committee may ask of the witnesses. And we ask that you respond as efficiently as possible.

And then finally, I will say the witnesses are excused, and the hearing is now adjourned.

[Whereupon, at 12:24 p.m., the Committee was adjourned.]

APPENDIX I: ANSWERS TO POST-HEARING QUESTIONS

Answers to Post-Hearing Questions
Responses by Dr. Richard B. Alley

Answers for the Record to Member Questions
House Committee on Science, Space and Technology

From the hearing
Earth's Thermometers: Glacial and Ice Sheet Melt in a Changing Climate
United States House of Representatives
Room 2318 of the Rayburn House Office Building
10:00 a.m., July 11, 2019

Dr Richard B. Alley*
Pennsylvania State University

*Any opinions, findings, conclusions, or recommendations expressed in this publication are those of the author and do not necessarily reflect those of the Pennsylvania State University, the Intergovernmental Panel on Climate Change, the National Academy of Sciences, or other organizations.

Submitted by Chairwoman Eddie Bernice Johnson

1. What is the biggest knowledge gap that is essential to fill to improve our projections of future sea level rise due to glacial melt?
   - The short answer, in my opinion, is that the biggest knowledge gap is the lack of understanding needed to predict possible "tipping point" behavior of Thwaites Glacier, West Antarctica and other such outlets that could rapidly and greatly raise sea level.
   - The longer answer is that actionable results are most likely to be supplied from sustained, integrated research that addresses the complex range of issues required to project future sea-level rise due to glacial melt, including efforts to maintain and expand participation by young scientists.

My written testimony briefly outlined the possible tipping-point behavior that could greatly and rapidly raise sea level. An international effort, especially funded by the US National Science Foundation (NSF) and the UK National Environment Research Council (NERC), with contributions from other nations and agencies, is addressing this issue, but in my opinion is small compared to the scale of the challenge and the potential cost of the sea-level rise if Thwaites Glacier collapse proceeds rapidly. I am happy to provide extensive information on this topic if desired, with pointers to broad-based assessments of key unknowns and ways to address them, with as much technical detail as needed. The Scambos et al. (2017) paper cited below is a useful starting point. Simply put, the research community has considerable understanding of the known and unknown unknowns that stand between us and the ability to project whether sea-level rise under strong warming will be two feet, or three feet or four, or more than ten feet by the end of the century, the community is working hard with the available resources, but resource limitations are still reducing our ability to move forward. (I know this is self-serving-I am a member of that community, and various colleagues and I personally have benefitted from the resources already made available-but I believe my assessment is accurate.)

More broadly, though, the key question of ice-sheet tipping points is inextricably coupled to a larger research framework. Allow me to make a few points, which I believe reflect the insights of the community, although I am presenting them as my opinions as an individual:

1. The solutions involve helping young people enter and remain in our field. I am privileged to teach, and so I get to speak to many young people who are interested in helping their country and the world prepare for the future. These bright students can go in any direction they want-they can conduct seismic surveys of great value to oil companies, sift through "big data" in ways that interest tech firms and banking companies, write computer code that could be used almost anywhere in the economy, and more, while communicating accurately and engagingly. Most of these young people will go to industry, finance, and other parts of the economy, but a few of them want to work in the climate-energy-economy nexus to extend our knowledge of climate change and use that knowledge to help people, even though these students expect to take a pay cut compared to their other options. But, these students express grave doubts about the wisdom of studying climate; they are not confident that the US government and other funding sources are sufficiently committed to this area to support a career. (These reports are anecdotal, and collected off-the-record so I cannot quote names and dates for you, but I have heard these worries repeatedly.)

   Our family just toured the amazing "Destination Moon: The Apollo II Mission" exhibit from the Smithsonian Institution in partnership with The Museum of Flight in Seattle, a wonderful testament to problem-solving and exploration. Many of the displays highlighted the unsung heroes at home who were essential to the success of the moon landing, including those in Houston: "When Apollo missions began flying in 1968, Mission Control had become a magnet for the best and brightest young engineers in the country." Solving the problems of climate, energy and economy requires getting a few more oftoday's best and brightest scientists and engineers to help us now, not scaring them away.

2. The solutions are broad-based. NSF is the US lead on studies of Thwaites Glacier now, but the observational and scientific contributions of NASA are essential, ice-sheet changes are tightly coupled to atmospheric and oceanic climate in ways that must involve NOAA, the modeling capability of DOE is increasingly important, the US Army Corps of Engineers has long contributed through their Cold Regions Research and Engineering Lab, the Coast Guard icebreakers have been essential while the 109th Airlift Wing of the New York Air National Guard provides heavy-lift aircraft for deep-field access on ice sheets, and contributions of USGS, the National Park Service, and others are important and can grow. Philanthropic funding has helped us in many ways, and we are deeply grateful for that, but does not in any way replace the broad-based, publicly accountable role of the US Government.
3. The known-knowns are important as well as the less-known or unknown unknowns of tipping-point behavior. If tipping points are not crossed, important uncertainties still remain about mountain glaciers, surface melting in Greenland, and more. The question, below, from Congresswoman Lofgren highlights possible linkages between ice melt and changes in ocean circulation, but those in tum will affect the pattern of sea-level rise by changing the size and location of the places that winds and currents pile up water against the shore or move water away. Coastal planners surely want to know whether they face more than ten feet of sea-level rise or "just" three feet, but they also are very interested in knowing whether that three feet could be two feet or four feet.
4. Research addressing the known-unknowns contributes to understanding the tipping points of the big ice sheets. For example, those of us who started out working on the big ice sheets have relied on work by many people including Professor Pfeffer on the

behavior of Alaskan tidewater glaciers; knowledge of the history and mechanisms of rapid ice retreat in coastal locations informs our understanding of possible rapid retreat in Antarctica. For various reasons, iceberg calving is often best studied on Greenland's ice now. Better ways to combine data and models, and continuing improvement in data collected through and beneath the ice, are frontiers where much essential progress can be made.

Again expressing my opinion, I believe that if you pressed most members of the community to pick a single research target, they would choose further research on ice-sheet tipping points. But, I also believe that they would make that recommendation without enthusiasm, because they understand the broad-based nature of the problem and the integrated approach needed to find solutions, surely involving helping students as well as satellites and other research tools.

Scarnbos, T., Bell, R. E., Alley, R. B., Anandakrishnan, S., Bromwich, D. H., Brunt, K., Christianson, K., Creyts, T., Das, S., DeConto, R., Dutrieux, P., Fricker, H. A., Holland, D., MacGregor, J., Medley, B., Nicholas, J. P., Pollard, D., Siegfried, M. R., Smith, A. M., Steig, E. J., Trusel, L. D., Vaughan, D. G. & Yager, P. L. (2017). How Much, How Fast?: A Review and Science Plan for Research on the Instability of Antarctica's Thwaites Glacier in the 21st century. *Global and Planetary Change, 153*, 16-34.

Submitted by Congresswoman Zoe Lofgren (AC-19)

1. Studies suggest that slower Atlantic Ocean circulation would cause colder winters and hotter summers in Europe, change rainfall patterns in the tropics, and pool warmer water along the U.S. East Coast which would contribute to sea level rise and more extreme storms. We've already seen how warmer water off the East Coast fueled extremely destructive hurricanes like Hurricane Maria and Superstorm Sandy.

(a) How do melting glaciers affect ocean circulation, and to what extent is research exploring this feedback loop?

The biggest role of melting glaciers is to raise sea level, flooding coasts and also potentially increasing storm surges by putting deeper water closer to buildings along the shore. But, this question does correctly identify a possible role for meltwater in ocean circulation. The most important sites for such influence are in or near the polar regions, in the north Atlantic and in the Southern Ocean, although changes are possible elsewhere.

Ocean circulation is driven in part by small differences in the saltiness of the water affecting the water density, so adding meltwater can indeed affect circulation. The best-known concerns point at the likelihood that freshwater added to the North Atlantic is slowing the rate at which water sinks to the depths there, affecting weather, climate, ecosystems and more. We are moderately confident that there will not be large, rapid and catastrophic changes, but important uncertainties remain despite rapid progress. Key monitoring and research are in place moving forward, but the ocean and the ice are changing rapidly and may get ahead of our learning, so I believe that additional research could be justified strongly, involving data collection today, model-data integration, and further work on the history of this complex and critical system.

Recent attention has focused on related changes in the Southern Ocean. Observational advances have shown us how much we don't know, and more clearly illustrated how tightly coupled the ice, ocean, air and ecosystems are in that vast ocean around Antarctica. Accelerating flow and melting of the Antarctic ice sheet dumping water and icebergs into the Southern Ocean could actually slow warming or cause cooling at the surface far to the south, but allow warmer waters beneath to reach the ice sheet and drive faster melting and ice loss, perhaps crossing the threshold for a tipping point. Deep uncertainty is attached to this possibility, with much research needed to project how this will evolve in a warming world. Again, sustained observations of ongoing changes, data-model integration, and studies of climate history can move understanding forward.

(b) Dr. Alley, in your testimony, you discuss how ice sheet melt is causing Earth's rotation to slow down very slightly and likens it to a spinning ice skater sticking out her arms. What are the implications of a slower rotation of the Earth?

The direct implications of changing Earth rotation are relatively small, but the science is important. Experts maintaining precise time must relate our noon-midnight timekeeping on a world with slowing rotation to the exact passage of time, and so must account for changes in the length of the day, but the technical means of doing so are well-known and functional. The detailed pattern of sea-level rise is actually affected a little by the changes in Earth's rotation, and that effect must be included in projecting future sea-level rise; this is broadly functional, although ongoing research can improve the results.

The point I hoped to make by introducing this topic in my testimony is related to just how much effort the scientific community puts into getting the right answers, how strongly interwoven the threads of knowledge are, and thus how reliable the results are that the community brings forward through assessments to policymakers such as yourself and your Committee. Measurements of growth or shrinkage of ice sheets are made by repeatedly weighing the ice sheets using satellites, by watching the surface elevation of the ice sheets change, and by comparing the snow added and the ice lost by melting and iceberg calving. These measurements agree within the known uncertainties of the measurements, and together give very high confidence that the ice sheets are losing mass. The water from this ice loss goes into the ocean, and the rise of sea level is what is expected from loss of mountain glaciers and ice sheets together with expansion of the ocean as it warms, providing a further check on the different measurement techniques. But, these also make predictions about changes in Earth's rotation, and when scientists check, these predictions are borne out-the different measurement techniques agree with each other, with the measured sea-level rise, and with the changes in Earth's rotation. We have even looked back in time at the record of eclipses as observed by past civilizations because the changing length of day affects those observations, confirming our understanding of the history of sea-level rise and the recent acceleration as human-caused warming melts ice.

You may occasionally hear some critic dismissing scientific knowledge as being uncertain and unreliable. Earth's rotational change is one of many examples showing that the scientific results are confident and reliable. When the scientific community brings forward to policymakers assessed results through the National Academies of Sciences, Engineering and Medicine, or the Intergovernmental Panel on Climate Change, those results rest on an interwoven fabric of evidence, such that even if a weakness were found in one thread, the fabric would remain intact. The occasional caricature of science as resting on one study or one investigator is simply wrong. If Einstein were erased from history our understanding of relativity would not change because so much knowledge has accumulated; we celebrate the pioneers, but rely on their insights only after extensive testing followed by assessment. The reliability of climate science never rests on one study or one scientist but instead on multiple strong lines of evidence.

Responses by Dr. Robin E. Bell

Response to Supplementary Questions: Glacier Melt Hearing Prof. Robin Elizabeth Bell August 8, 2019

Chairwoman Johnson Question 1*You have described how the work of understanding changing ice and glacier melt is spread across multiple agencies. Is this effective and if not what changes would you recommend?*

The federal work on changing ice and glacier melt and its impacts is spread across multiple agencies. The scope of this impressive work ranges from fundamental research on how ice sheets work (NSF and NASA), to earth observations from space of how the ice is changing (NASA), to building models that will forecast future change (NSF, DOE, NASA). But the changing ice does not stay at the poles as documented by measurements spanning over a century (NOAA) and decades (NASA). The impacts of glacial melt and changing ice are being observed and quantified along our changing coastlines by the USGS. Simultaneously states, counties, municipalities and many branches of the Federal government including the Army Corps of Engineers, National Park Service, USDA, and Department of Defense, are dealing with the consequences. All these stakeholders are looking to the glacial scientific community for information and projections to help them plan. The National Climate Assessment provides a snapshot of sea level projections and some of what has been done in communities and is a very important resource, but there is no comprehensive plan on how our nation will respond to glacial melt and its impacts on communities.

Now is the time to act. We must develop a strategy to link the rich glacial melt research underway with the needs of the nation especially the agencies and communities now facing accelerating change. A plan on how to holistically bridge this gap is essential. The scientific community knows improving our projections of future ice will melt is essential but a gap still remains between the changing ice community and the changing coastline community. This gap is between global projections and what local communities need to build effective responses. Different communities have different needs. We must develop a plan to respond to changing sea level in the coming century even as we work to constrain the actual numbers. Preparing for sea level rise is like preparing for an earthquake. Our communities benefit greatly from planning before an earthquake even if we do not know exactly when the earthquake will come or how large it will be. Melting glaciers, changing ice and changing coastline require the same planning. I believe it is absolutely essential to create a better framework to prepare the change we know is coming. A major study by the National Academies of Sciences, Engineering, and Medicine linking effective preparations for changing sea level with the science of the changing ice, warming oceans and changing coastlines could help identify gaps in our knowledge and build synergies that will better prepare communities at risk. A study would encompass an inventory of the existing programs, strategies for response, and develop frameworks for communities, agencies and individuals to respond. This study would strengthen the connectivity among the experts working all the facets of the issue to those responsible for planning and mitigation efforts. Such a study would build on the recent National Climate Assessment and strengthen the next. A congressional recommendation for an Academy study to develop a robust response to changing ice and changing coastlines would be a major service to all.

Chairwoman Johnson Question 2 *Does the Notional Climate Assessment play a role in preparing our coastal communities for the impacts of glacial melt we can anticipate in the coming decades and if so what is that role?*

Every four years, the National Climate Assessment presents the state of the art of the science behind sea level rise in the climate science section and evaluates the impacts and adaptation both across sectors and for all the regions of the U.S. It is an illustration of how the science and our communities can work together to quantify the risks to changing climate and share impacts and responses. This structured national process is as essential to planning for our future as is the census- the Founding Father's just has not considered climate change yet. With this structured review of the science agencies, communities, and individuals have an opportunity to understand what is happening, how it is impacting communities and what sort of solutions are being developed. A remarkable resource, the National Climate Assessment plays an important role in making it identifying the impacts of climate with detail and clarity.

Chairwoman Johnson Question 3 *What is the biggest knowledge gap that is essential to improve our projections of future sea level rise due to glacial melt?*

The biggest gap that is essential to improve our predictions of future sea level rise is our knowledge of how Antarctica will change in the future. We know where it is changing most quickly, and we see recently accelerating hot spots. What we do not know is how the warming ocean is reaching the ice -the trigger that has started the speedup - and we do not know what will happen when the air warms up and the surface of Antarctica begins to look like Greenland. Antarctica impacts the most people in the U.S. and globally, and it the biggest wild card in our sea level rise projections. Addressing these questions will require major focused fieldwork and probably large scale international collaborations. Antarctica is simply hard to work in. It is big, cold and remote. Even though it is warming quickly it is still inhospitable and a challenging place to work.

Chairwoman Johnson Question 4 *Given the understanding that sea levels will be different for each community due to gravitation fingerprinting and other local factors, how do you think we can make personal for communities the connections between global changes in ice with the changes we will see here in the U.S. and around the globe?*

When I began to realize how complex sea level rise it struck me that it was essential to move past making isolated statements of sea level rise from individual ice sheets or even a global mean. Who knew how complex and beautiful our planet could be? Communities need to know everything from where the ice is melting, to whether their region is sinking or rising and how the ocean is warming. An effective approach can frame how sea level has changed in the past, considering what range of change a community will see in the future and will look at the assets close to sea level. It is essential to convey that we can develop a plan, work to minimize the change and mitigate the uncertainties and surprises around sea level rise.

Chairwoman Johnson Question 5 *In your research field, what is the importance of convergent research? How will it help the U.S. address glacial melt and foster resilience in our coastal communities?*

Convergent science emerged from the biomedical science where advances on complex problems are only possible with a broad team of experts working closely together. The ROSETTA-Ice team working on Antarctica's Ross Ice Shelf, which I lead, illustrates the power of the convergent approach. The ROSETTA-Ice team of geologists, geophysicists, oceanographers, atmospheric scientists, engineers and glaciologists produced novel insights because it brought such diverse expertise to the table. The team discovered that the geologic framework of the Ross Sea is preventing warm ocean water from reaching the grounding line of the ice shelf but discovered the sensitivity to change is at the ice shelf front where local winds can bring in heat. The next 100 years must focus on convergent collaborations to advance our understanding on how ice sheets change. This type of science is hard. It is easier to look at a problem from the safe base of your knowledge but reaching across disciplines can be rewarding and provide key insights. Our earth and the ice our planet does not care about disciplines. More broadly convergent changing ice research will link the changing ice in the poles to our impacted coastlines and communities. Connecting our understanding of changing ice to changing coastlines is beginning but must expand in the coming decade. Each community will need an individual projection that includes ocean warming, melting ice, changing land levels and coastal dynamics. It is essential that we educate next generation convergent changing ice and changing coastlines scientists. This convergent workforce will actively engage with the entire breadth of scientists, engineers, social scientists and policymakers who are essential to providing communities with the information they need to plan their infrastructure investments and land use in the coming century

Congresswoman Hill Question 1a Dr. Bell can you expand on your role as Chair of the National Academies review of the National Climate Assessment?

Peer-review of science is one of the gold standards of the modern scientific process. I served as the chair of a panel of 16 experts who, through the National Academies of Sciences, peer reviewed the National Climate Assessment. I lead the committee through the process of reviewing all the chapters and identifying issues within the document. As we reviewed the report we considered whether the information was broadly accurate and represented the current state of understanding at a level that would inform decision-making and be accessible to a broad, non technical audience.

The National Climate Assessment is a key document that communicates how climate change is occurring in our backyards and, that together with strategies informed by research, our communities can be resilient. In addition to sea level rise, the report covers human health and community well-being, the built environment, businesses and economies, ecosystems and natural resources. Hundreds of experts from federal, state, and local governments, academia, non-government organizations, and the private sector assembled the report. The Assessment process also gathers from community engagement events and public comment. By bringing together such a deep array of perspectives on climate change impacts and resilience, the effort represents a tremendous large-scale collaboration. Because the climate change report highlights adaptation and mitigation efforts, it serves as a valuable resource for many. We delivered our report (that itself was peer reviewed through the National Academy process) to the National Climate Assessment team and they addressed the issues we raised. We did ask they increase discussion of adaptation and mitigation efforts emerging as these are critical to communities as they address the change.

Reading the 1500 pages on the impact of the climate on our country, from the Virgin Islands to the Pacific territories, gave me and the committee a profound understanding of the scope of climate change that our nation is already experiencing. Melting glaciers and changing sea level is one of the strongest signals of the change across the U.S. but the change is right in our backyards.

Congresswoman Hill Question 1b What is the importance of this document to understanding the consensus in this field?

The National Climate Assessment provides both the state-of-the art science across the U.S. and its territories and a detailed tracing of the source of the information. The National Climate Assessment presentation of the data (such as temperature change) is at such a detailed level it is possible to see how much temperature has changed in individual communities. Some news outlets have developed useful ways to look up change in their community by entering a zip code. Here we are seeing the sharing of the NCAA again and again. The document not only demonstrates the consensus of scientific knowledge but also allows citizens to see where they and their communities fit into the patchwork of change and impacts

Congresswoman Hill Question 1c What is its importance to our understanding of climate science and its effects on society more broadly?

The National Climate Assessment is a very powerful process that brings together climate experts, the state of the art data and projections to communities who are both witnessing and planning for this change. Breaking out the impacts and adaptations by region is a powerful tool for communities and citizens to understand how climate change is evolving around them. Often individuals have a difficult time visualizing the change greenhouse gases (invisible) and a relatively small temperature rise can mean. However, the evidence of the longer growing seasons-up to 40 days longer in California-is a powerful way to advance the understanding of climate impacts. People's daffodils and crocuses are coming up sooner in the spring and their tomatoes and basil are lasting longer in the fall. Seed companies have remapped hardiness zones in their catalogues. Climate change is effecting what fish people pull from the ocean, because cold loving fish shift north in the Atlantic while in the Gulf of Mexico cold loving fish are being replaced by warmer (less tasty) species. The National Climate Assessment documents how climate change is impacting local communities from the plants they grow to the take at the end of their fishing lines. No other efforts document local effects comprehensively. This process enables communities to understand the change and begin to develop resilience plans.

Responses by Dr. Twila A. Moon Questions for the record for the House Committee on Science, Space, and Technology for the hearing on Earth's Thermometers: Glacial and Ice Sheet Melt in a Changing Climate provided on August 2, 2019 by Twila Moon, PhD Research Scientist, National Snow and Ice Data Center, Cooperative Institute for Research in Environmental Sciences, University of Colorado Boulder

Question 1) Given the importance of glacial meltwater as a freshwater supply for certain animals and plants, such as salmon in the Pacific Northwest, what are the implications of increased glacial melt rates to these species and ecosystems?

Response:

As the climate warms, glacier melt is increasing. In a glacierized basin, glacier melt makes up some percentage of the stream and river flow, with that contribution changing through the seasons and depending on weather conditions. As the glaciers begins to melt more, there can initially be an increase in glacier melt input (potentially lasting across many years). However, as the glacier area continues to decline, there is less ice remaining to melt and glacier runoff will decrease. The result is that ecosystems must respond to complex variability that can include both increases and decreases in glacier meltwater input, influencing for example temperature, turbidity, and total stream flow amount.

The implications of increased glacial melt for salmon, other species, and the ecosystems they depend on varies by location and in some cases is not known at all. For example, in the Pacific coastal temperate rainforest *of* Alaska, scientists foresee possible positive and negative consequences. Streams in this area with particularly high glacier meltwater content can be too cold for salmon rearing and overwintering. So reduction in glacier melt in some streams may make them better salmon habitat. However, water temperatures that are too warm in the summer because glacier melt input has dropped too low may hurt salmon productivity. It may be that different salmon species are helped or hurt by glacier loss in specific locations. Further discussion of the Pacific coastal temperate rainforest ecosystem and connections between ecosystem and glacier loss can be found in O'Neel et al. (2015). For many other locations, the impacts of losing glacier ice are more clearly negative. For example, the meltwater stonefly *Lednia tumana,* which is endemic to Waterton-Giacier International Peace Park in Canada and the U.S. and has been petitioned for listing under the U.S. Endangered Species Act, lives in a restricted range near glaciers, snowfield, and springs, and changes in habitat from glacier loss-including increasing water temperatures, changes in the timing and amount of glacial meltwater input, changes in dissolved oxygen levels, and alteration of stream flow character are expected to negatively impact this species (e.g., Muhlfeld et al. 2011). Reductions in ecosystem health in the base of the food chain is likely to negatively impact other dependent species.

Overall, however, it is my assessment that research into links between glacier loss and dependent plants, animals, and ecosystems is limited. This is particularly worrisome given that glaciers are changing rapidly in parallel with other potentially large changes, such as increased temperature and changing precipitation. It is likely that multiple climate impacts will be at play in any given ecosystem, and it will be important for research to consider connections across human, biological, and physical systems.

References

Muhlfeld, C. C., Giersch, J. J., Hauer, F. R., Pederson, G. T., Luikart, G., Peterson, D. P., Downs, C. C. & Fagre, D. B. (2011). Climate change links fate of glaciers and an endemic alpine invertebrate, *Climatic Change*, *106*(2), 337-345,doi:10.1007/s10584-011-0057-1.

O'Neel, S., et al. (2015). Icefield-to-Ocean Linkages across the Northern Pacific Coastal Temperate Rainforest Ecosystem, *BioScience*, *65*(5), 499-512,doi:10.1093/biosci/biv027.

Question 2) Local communities that rely on glacial streams for water for drinking, industry, or agriculture may be impacted by drought in the future as glacial runoff diminishes. What are the impacts of melting glaciers on local and global drinking water supplies in the U.S. and around the world?

Response:

There is no doubt that worldwide loss of glacier ice will impact water supplies around the globe. First, as mentioned above, it is important to recognize that increasing glacier melt can produce both an initial bump in water availability and a subsequent, long-lasting decline. Glacier meltwater is also most available during summer, when air temperatures are above freezing, and is therefore particularly important as a water source from spring through fall.

Depending on the local or regional environment, glacier melt can be vital as a source for drinking water, irrigation and agricultural water, and hydropower and other industry. For example, in south Asia (including Pakistan, Afghanistan, Tajikistan, Turkmenistan, Uzbekistan, and Kyrgyzstan), 800 million people depend on water resources served by Himalayan glaciers. The role of glacier melt in this area is particularly strong during drought years, when glacier melt offsets the low water input from rain. Projected future declines in glacier area across the Himalaya will impact water availability, adding another stressor to a region that already grapples with geopolitical tensions (for a full discussion of this case study see Pritchard (2019)). These challenges are not unique to south Asia (e.g., Carrey et al. 2017). In South America, glaciers provide valuable water resources for many countries. In Bolivia, for example, the capital city of La Paz is growing rapidly, but is also dependent on water resources that include substantial glacier meltwater. During 1963-2006, glaciers contributed 14% of La Paz's water resources in the wet season and 27% in the dry season. While increased glacier melt during the latter period of this 1963-2006 record had so far offset the loss of 50% of the glacier area, eventual complete loss of the region's glaciers will cause a 9% reduction in total runoff during wet seasons and 24% reduction during the dry season (for further discussion on this case study see Soruco et al. (2015)). Even in the United States, glacier meltwater can be an important local source of runoff, for example in the Pacific Northwest (Nolin et al. 2010). It is my personal opinion, however, that the wide-ranging impacts of glacier loss within the United States are understudied.

References

Carey, M., Molden, O. C., Rasmussen, M. B., Jackson, M., Nolin, A. W. & Mark, B. G. (2017). Impacts of Glacier Recession and Declining Meltwater on Mountain Societies, *Annals of the American Association of Geographers*, *107*(2), 350-359, doi:10.1080/ 24694452.2016. 1243039.

Nolin, A. W., Phillippe, J., Jefferson, A. & Lewis, S. L. (2010). Present-day and future contributions of glacier runoff to summertime flows in a Pacific Northwest watershed: Implications for water resources, *Water Resour. Res.*, *46*, W12509, doi:10.1029/2009WR008968.

Pritchard, H. D. (2019). Asia's shrinking glaciers protect large populations from drought stress, *Nature*, 1-20, doi:10.1038/s41586-019-1240-1.

Soruco, A., Vincent, C., Rabatel, A., Francou, B., Thibert, E., Sicart, J. & Condom, T. (2015). Contribution of glacier runoff to water resources of La Paz city, Bolivia (16° S), *Annals of Glaciology*, *56*(70), 147-154, doi:10.3189/2015AoG70A001.

Responses by Dr. Gabriel J. Wolken, The Impacts of Glacier Change in Alaska, Testimony of Gabriel J. Wolken, Alaska Division of Geological & Geophysical Surveys and University of Alaska Fairbanks for the hearing entitled Earth's Thermometers: Glacial and Ice Sheet Melt in a Changing Climate Before the Committee on Science, Space, and Technology United States, House of Representatives, 11 July 2019

Questions for the Record

1. Given that Alaska has the highest number of glaciers out of any state in the United States and those glaciers are melting at substantial rates, what are the projected economic impacts of glacial melt to Alaska's economy, and does the state of Alaska have any adaptation or mitigation strategies for addressing loss of its glaciers?

Response:

The rapid and sustained melting of glaciers in Alaska will negatively impact Alaska's economy.

As glaciers in Alaska continue to lose mass, a rise in sea level will lead to more lowland flooding and coastal erosion that damages or destroys infrastructure, threatens public safety, security and livelihoods, and increases the financial pressure on economies at all levels.

By the end of this century, glaciers in Alaska are projected to lose about 30-50% of their current total mass. This will have a large negative impact on runoff volume, modifying the chemical composition and decreasing the temperature of many of the streams that support key fisheries, including salmon, and the billion dollar fishing industry on which many Alaskans depend. Changes in runoff will also affect hydropower production. As glacier mass loss continues, the storage capacity of snow and ice will eventually diminish, which is projected to have significant impacts on future water resource availability, even in basins with low glacier coverage. This will lead to an economic burden on the communities in Alaska that currently rely on hydropower generation from glacierized catchments.

Alaska's tourism industry may also be negatively impacted by continued glacier mass loss. Tourism in Alaska is a key economic sector that supports one out of every ten jobs in the state, delivers 1.5 billion dollars in labor income, and has an economic impact of 4.5 billion dollars. Recent glacier retreat has already been significant in Alaska, making it more difficult for tourists to access glaciers using Alaska's road system, and tourism operators are being forced to modify the manner in which their clients experience glaciers.

Few adaptation and mitigation strategies have been implemented, or developed for addressing the impacts of glacier melt in Alaska. The State of Alaska, and some municipalities, have updated the language in their hazard mitigation plans to include glacier-related hazards associated with continued glacier mass loss, and exposure to extreme runoff events. Some communities are experiencing the impacts in a very direct way. For instance, the City of Valdez has spent nearly 6 million dollars in adaptation responses to glacier lake outburst floods resulting from the rapid retreat of Valdez Glacier. In other parts of Alaska, lowland flooding and coastal erosion, resulting in part from rising sea level, are forcing communities and governments to evaluate their adaptive capacities to these rapidly changing conditions; in some locations, relocation is the only option.

Alaska will continue to experience the effects of climate change more than most states in the U.S. because of its high northern latitude location and abundance of snow, ice and permafrost. Alaska is warming at twice the rate as the rest of the U.S., and rapid changes in the snow, ice, and permafrost will increase the occurrence, and likely the magnitude, of hazards throughout the state (e.g., avalanches, floods, erosion, slope instability, glacier collapses, and glacier lake outburst floods). If these hazards are not properly assessed and monitored, and if the appropriate adaptive measures are not considered, these cryosphere hazards will have a damaging effect on Alaska's communities and infrastructure, as well as on the security and livelihoods of Alaskans.

Responses by Dr. W. Tad Pfeffer

W.T. Pfeffer Testimony to US House of Representatives Committee on Science, Space, and Technology 11 July 2019

Response to question from Rep. J. McNerney (CA-09): "You mentioned in your testimony the fact that the Arctic gyre is known to trap air. Can you describe how this phenomenon would impact climate interventions, such as the injection of cloud aerosols, in the Arctic, and vice versa?"

My expertise does not extend to arctic climate dynamics and especially not to questions of geoengineering, but I can provide some background to your question and suggest other researchers who are expert in these areas.

The presence of a circumpolar closed low-pressure system has been known since the 19th century, but the presence and significance of an upper-level (upper tropospheric- stratospheric) cyclonic circulation was not detected until the mid-1950s when high-altitude weather reconnaissance missions were flown from Alaska into the high Arctic (Shaw, 1995). Later analysis of data acquired during these flights help to identify the mechanism responsible for "Arctic Haze," episodes of reduced visibility in the arctic first described early in the mid-18th century, at the outset of the industrial revolution. This haze is caused by scattering and attenuation of light by small aerosol particles, and from the work of Shaw and others has been shown to originate predominantly from high-latitude coal burning industries in Eurasia. The 1995 Shaw paper (PDF copies of this and other papers accompanies this response) gives a good summary overview of the Arctic Haze phenomenon and what it reveals about arctic atmospheric circulation.

Regarding the climate-change implications of arctic atmospheric circulation, probably the most important aspect is the changing strength, position, and mobility of the upper level arctic jet and its effects on mid latitude weather and climate.

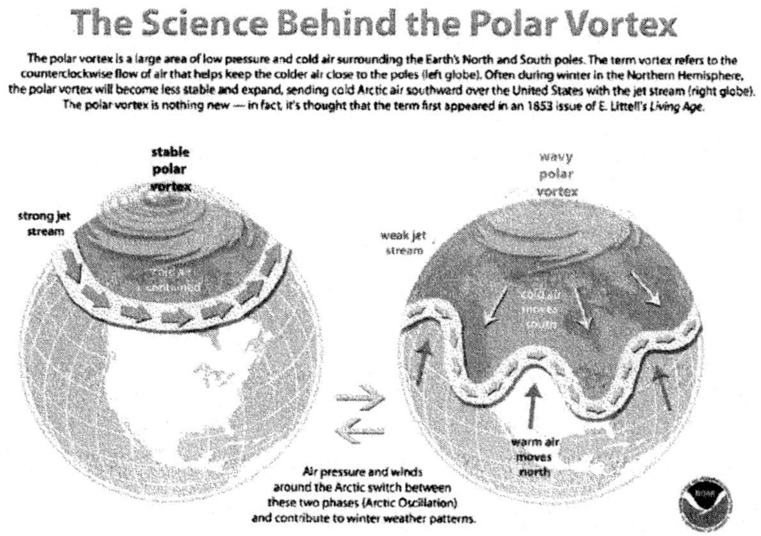

The arctic jet is an upper level band of strong westerly wind that separates the cold arctic air mass to the north from the warmer mid-latitude air to the south (see the accompanying graphic). The jet does not move in a simple circular pattern from west to east at a fixed latitude (as shown in the "stable polar vortex" in the graphic), but wanders in a looping pattern (as shown in the "wavy polar vortex" in the graphic).

These looping waves tend to wander about fixed average positions on various time scales ranging from synoptic (i.e., weather systems) to seasonal (winter vs. summer), but also appear to be moving and weakening in response to climate change forcings. Seasonal changes in the position of the polar jet cause familiar regular changes in seasonal climate (e.g., in New England, the transition from hot/humid weather in July to cool/dry weather in August), while snoptic-scale changes cause other familiar phenomena like cold incursions into the Midwest in the winter (the "Alberta Clipper"). Recent anomalous events like the January 2019 cold outbreak in central USA are manifestations of a polar jet changing in new and unexpected patterns. Climate change appears to be altering the dynamics of the arctic jet (and arctic circulation generally), and more anomalous weather and seasonal patterns may be expected in the future. I have included some papers covering this issue (Chen et al, 2914; Zhang et al; 2016), and recommend Dr. Jennifer Francis (Woods Hole Institute) as an expert in this subject.

I can offer no general information or advice regarding geoengineering and arctic climate beyond noting that geoengineering approaches are being discussed (see Caldeira and Wood, 2008; and Dykema et al, 2014). I recommend Prof. Alan Robock (Rutgers University) as an expert in this subject.

Further reading:

Arctic Haze entry in Wikipedia (https:ljen.wikipedia.org/wiki/Arctic haze)

Recommended Experts:

Arctic climate and polar vortex: Dr. Jennifer Francis, Woods Hole Research Center, Woods Hole, MA (jfrancis@whrc.org)

Geoengineering: Prof. Alan Robock..Rutgers University (robock@envsci.rutgers.edu)

References

Caldeira, K. & Wood, L. (2008). Global and Arctic climate engineering: numerical model studies. *Philosophical Transactions of the Royal Society A: Mathematical, Physical and Engineering Sciences*, *366*(1882), 4039-4056.

Cohen, J., Screen, J. A., Furtado, J. C., Barlow, M., Whittleston, D., Coumou, D. & Jones, J. (2014). Recent Arctic amplification and extreme mid-latitude weather. *Nature geoscience*, *7*(9), 627.

Dykema, J. A., Keith, D. W., Anderson, J. G. & Weisenstein, D. (2014). Stratospheric controlled perturbation experiment: a small-scale experiment to improve understanding of the risks of solar geoengineering. *Philosophical Transactions of the Royal Society A: Mathematical, Physical and Engineering Sciences*, *372*(2031), 20140059.

Francis, J. A. & Vavrus, S. J. (2012). Evidence linking Arctic amplification to extreme weather in mid latitudes. *Geophysical Research Letters*, *39*(6).

Shaw, G. E. (1995). The Arctic haze phenomenon. *Bulletin of the Amencan Meteorological Society*, *76*(12), 2403-2414.

Zhang, J., Tian, W., Chipperfield, M. P., Xie, F. & Huang, J. (2016). Persistent shift of the Arctic polar vortex towards the Eurasian continent in recent decades. *Nature Climate Change*, *6*(12), 1094.

APPENDIX II: ADDITIONAL MATERIAL FOR THE RECORD

Letter Submitted by Representative Suzanne Bonamici

AGU PUBLICATIONS

Geophysical Research Letters

Dissolved Black Carbon in the Global cryosphere: Concentrations and Chemical Signatures

Alia L. Khan[1,2], Sasha Wagner[3,4], Rudolf Jaffe[3], Peng Xian[5], Mark Williams[6], Richard Armstrong[1], and Diane McKnight[2]

[1]Department of Civil and Environmental Engineering and the Institute of Arctic and Alpine Research, University of Colorado Boulder, Boulder, Colorado, USA, [2]National Snow and Ice Data Center and Cooperative Institute for Research in Environmental Sciences, University of Colorado Boulder, Boulder, Colorado, USA, [3]Southeast Environmental Research Center and Department of Chemistry and Biochemistry, Florida International University, Miami, Florida, USA, [4]Skidaway Institute of Oceanography, Marine Sciences Department, University of Georgia, Savannah, Georgia, USA, [5]Aerosol and Radiation Section of the Marine Meteorology Division, Naval Research Laboratory, Monterey, California, USA, [6]Department of Geography and the Institute of Arctic and Alpine Research, University of Colorado Boulder, Boulder, Colorado, USA

Abstract

Black carbon (BC) is derived from the incomplete combustion of biomass and fossil fuels and can enhance glacial recession when deposited on snow and ice surfaces. Here we explore the influence of environmental conditions and the proximity to anthropogenic sources on the concentration and composition of dissolved black carbon (DBC), as measured by benzenepolycaroxylic acid (BPCA) markers, across snow, fakes, and streams from the global cryosphere. Data are presented from Antarctica, the Arctic, and high alpine regions of the Himalayas, Rockies, Andes, and Alps. DBC concentrations spanned from 0.62 μg/L to 170 μg/L. The median and (2.5, 97.5) quantiles in the pristine samples were 1.8 μg/L (0.62, 12), and nonpristine samples were 21 μg/L (1.6, 170). DBC is susceptible to photodegradation when exposed to solar radiation. This process leads to a less condensed BPCA signature. In general, DBC across the data set was composed of less polycondensed DBC. However, DBC from the Greenland Ice Sheet (GRIS) had a highly condensed BPCA molecular signature. This could be due to recent deposition of BC from Canadian wildfires. Variation in DBC appears to be driven by a combination of photochemical processing and the source combustion conditions under which the DBC was formed. Overall, DBC was found to persist across the global cryosphere in both pristine and nonpristine snow and surface waters. The high concentration of DBC measured in supraglacial melt on the GRIS suggests that DBC can be mobilized across ice surfaces. This is significant because these processes may jointly exacerbate surface albedo reduction in the cryosphere.

Plain Language Summary

Here we present dissolved black carbon (DBC) results for snow and glacial melt systems in Antarctica, the Arctic, and high alpine regions of the Himalayas, Rockies, Andes, and Alps. Across the global cryosphere, DBC composition appears to be a result of photochemical processes occurring en route in the atmosphere or in situ on the snow or ice surface, as well as the combustion conditions under which the DBC was formed. We show that samples from the Greenland Ice Sheet (GRIS) have a distinct molecular chemical signature, consistent with deposition of BC from Canadian wildfires occurring the week before sampling. The concentration range observed in this global cryosphere study indicates significant amounts of DBC persist in both pristine and human impacted snow and glacial meltwater. Our results are significant for understanding the controls on meltwater production from glaciers worldwide and the feedbacks between combustion sources, wildfires, and the global cryosphere. Wildfires are predicted to increase due to climate change, and albedo cannibalism is already influencing meltwater generation on the GRIS. Anticipated longer summer melt seasons as a result of climate change may result in longer durations between snowfalls, enhancing exposure of recalcitrant DBC on snow/ice surfaces, which could further exacerbate surface albedo reduction in the cryosphere.

Introduction

Many inherent challenges remain in quantifying and predicting melt of polar ice sheets and glaciers. In particular, the complex influence of light-absorbing aerosols is not well understood [Bond et al., 2013]. Biomass burning and fossil fuel combustion are both sources of black carbon (BC) to the cryosphere [Fellman et al., 2015]. Organic matter from these sources can be stored in glaciers for millennia [Hood et al., 2009, 2015; Stubbins et al., 2012a] and mobilized during glacial melting [Hodson, 2014], A study on the Greenland Ice Sheet (GRIS) attributed two widespread melting events (1889 and 2012) to BC produced from Northern Hemisphere wildfires [Keegan et al., 2014]. This attribution was based on higher BC and ammonia concentrations in ice lenses in firn cores, along with air mass back trajectory analysis; however, direct chemical characterization of the BC was not obtained. Once deposited, BC aerosols can be solubilized and transported in the aqueous phase as dissolved black carbon (DBQ [Dittmar,

2008]. In Alaskan glacier rivers, DBC was interpreted as being sourced from atmospheric deposition of anthropogenic combustion products [Ding et al., 2014a]. In Antarctica, DBC from wildfires in preindustrial eras has accumulated over millennia in the saline bottom waters of closed-basin lakes; the DBC in the bottom waters has a more polycondensed chemical signature in comparison to DBC in surface waters susceptible to influence from current local anthropogenic point sources of BC [Khan et al., 2016]. Therefore, atmospheric deposition of both natural and anthropogenic BC aerosols represents an important source of DBC to remote areas of the cryosphere.

DBC is quantified and characterized using the benzenepolycarboxylic add (BPCA) method, which produces individual BPCA molecular markers upon the oxidation of condensed aromatic structures [Dittmar, 2008]. The condensed aromaticity of DBC can be inferred from the relative proportion of produced BPCAs [Schneider et al., 2010; Abiven et al., 2011]. The degree of aromatic condensation for particulate BC is primarily a function of pyrolysis temperature, and links between BPCA composition and pyrogenic source are not always clear [e.g., Schneider et al., 2010; Mcbeath et al., 2013a, 2013b; Wiedenmeyer et al., 2015].1n addition, DBC molecular signatures are altered via photodegradation, either en route or in situ, changing BPCA composition and obscuring potential links to its pyrogenic source [Schneider et al., 2010; Ziolkowski and Druffel, 2010; Stubbins et al., 2012b; Ward et al., 2014]. Although BPCA composition cannot be used to directly infer the pyrogenic source of DBC, we can use them to investigate potential biogeochemical drivers of the quality of DBC which persists in cryospheric environments. In this study, we hypothesize that a more condensed DBC signature will be characteristic of samples for sites with freshly deposited BC and limited time for photodegradation, while DBC enrichment in less condensed aromatics will be characteristic for sites where the DBC may have been exposed to solar radiation for long durations of time either in the atmosphere or on the snow/ice surface. In terms of sites with freshly deposited BC, we investigated sites on the GRIS where collection occurred shortly after Canadian wildfires. We also investigated sites where continuous inputs of local anthropogenic sources of BC may lead to a fresher signature. In addition to the GRIS, an area known to be impacted by wildfire-derived pyrogenic carbon, we included other polar regions with high solar exposure, which are exposed to continuous sunlight almost 4 months per year, as well as high-altitude mountain sites. Sampling locations in this study are largely remote, experience extended lengths of sunlight exposure, and receive diverse inputs of atmospheric BC from both natural and anthropogenic combustion sources. As such, our overall hypothesis is that in the cryosphere, degradative processing during DBC transit in the environment, as well as the combustion conditions under which the DBC was formed, are important drivers of the DBC BPCA profile across the cryosphere.

Here we assess variation in DBC content and composition in samples that represent the typical global cryosphere, as well as sites with known local sources of DBC (Figure 1: map). Data are presented from glacial melt systems in Antarctica, the Arctic (GRIS and Svalbard), and high alpine regions of the Himalayas, Rocky Mountains, Andes, and Alps. Snow samples are from the Rocky and Andes Mountains, as well as the Arctic. Samples from the Norwegian Arctic were collected on Svalbard, including around an active coalmine and coal burning power plant, fueling the largest settlement, Longyearbyen. Sites are categorized as pristine" (P; >5 km from a fuel combustion source) and "nonpristine" (NP; <5 km from a fuel combustion source). To support the interpretation of the GRIS results, we determined biomass burning smoke aerosol optical thickness (AOT) and air mass transport from the Navy Aerosol Analysis and Prediction System (NAAPS) over several days in the week before the samples were collected.

Methods

Water samples were obtained that represent the diverse aquatic environments of the cryosphere. The lakes and streams sampled include supraglacial systems found on the surface of glaciers, systems fed directly by glacial melt and/or snow melt, as well as proglacial lakes, which form at the tongue of glaciers.

There are also surface water ponds included from the McMurdo Dry Valleys (MDVs), including DL-Hoare Pond (DLH Pond), latitude: –77.622901916504, longitude: 162.902999877930, which is adjacent to Lake Hoare: http://www.mcmlter.org/content/dirty-little-hoare-pond. Lake Hoare was named for physicist Ray Hoare, of the eighth Victoria University Expedition (1963-1964). Additional samples were collected at Cape Royds in Antarctica. Regionally, the samples are divided into Arctic, Antarctic, and Alpine. The Arctic data set is mostly composed of samples from the island of Svalbard in the Norwegian Arctic, as well as two samples from the GRIS, one fresh snow and one supraglacial melt. Antarctic samples were collected from the MDV and Cape Royds. Alpine samples are from snow and snowmelt in the Colorado Rocky Mountains, supraglacial melt from the Ngozumba Glacier, the longest glacier in Nepal located in the Gokyo Valley of the Nepalese Himalayas, snow and glacier fed streams in the central Chilean Andes, and a glacier-fed stream from the Mer de Glace in the French Alps.

Water and snow samples were collected in acid-rinsed and precombusted amber glass bottles or acid-rinsed high-density polyethylene (HDPE) bottles. Snow was melted at room temperature in precleaned acid-rinsed Nalgene buckets. All samples were filtered on precombusted glass fiber filters (Whatman GF/F, pore size: 0.7 m), acidified to pH = 2 with concentrated HCI (analytical grade), and stored at 4°C until analysis. For all samples, a 60 mL aliquot was poured into acid-rinsed and precombusted amber glass bottles and analyzed for dissolved organic carbon on a Shimadzu TOC-V-CSN with a detection limit of 0.07 mg C/L.

DBC (<0.7 μm) was measured using the BPCA method [Dittmar, 2008; Dittmar et al., 2008] and optimization for freshwater DBC [Ding et al., 2014b]. This method is based on the oxidation of polycondensed DBC structures to BPCAs with three to six carboxylic acid groups (B3CA to B6CA). The BPCA method was applied to the dissolved organic matter (DOM) isolated by solid-phase extraction (SPE).

First, a SPE technique [Dittmar, 2008; Dittmar et al., 2008] was applied to all samples using Bond Elut PPL cartridges, composed of a styrene divinylbenzene polymer solid phase. SPE efficiency for marine samples is typically ~45% and ~60% in fresh water [Dittmar et al., 2008]. In this study we measured SPE efficiency on eight samples (five Antarctic surface waters (37%, 34%, 46%, 50%, and 67%), a glacier-fed stream from the Andes (34%), and two snow sample (47% and 73%).The average efficiency of these eight samples was 48% ± 15%, with a range from 34% to 73%.

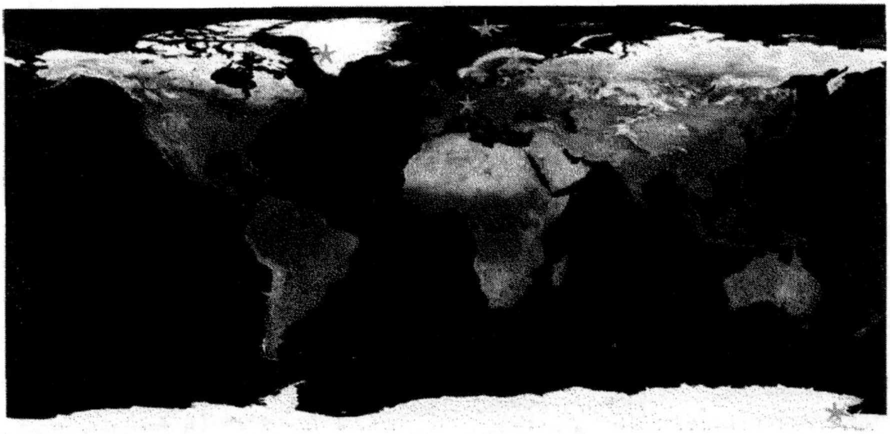

Figure 10. Map of global distribution of sample sites overlaid on a NASA Visible Earth image. Data are presented from glacial melt systems in Antarctica, the Arctic, and the Himalayas, Snow samples are from the Rocky and Andes Mountains, as well as the Arctic.

The PPL cartridges were transported frozen to Boulder, CO, and stored in a freezer. The PPL cartridges were transported to Miami, FL, and completely dried under ultrahigh-purity nitrogen gas. The PPL cartridges were then eluted with 10-20 mL of MeOH until the eluent was colorless. Aliquots of MeOH were then quantitatively transferred to 2 ml glass ampules and evaporated to dryness under a stream of ultrahigh-purity nitrogen.

The dried DOM extract was redissolved in concentrated nitric acid (65%) before ampules were sealed and oxidized in sealed glass ampules in a programmable oven for 6 h at 160°C in order to produce BPCAs [Ding et al., 2014b]. They were then dried in a sand bath under ultrahigh-purity nitrogen gas. BPCAs were then redissolved in the mobile phase buffer, separated, and quantified on a Sunfire C18 reversed phase column (3.5 μm, 2.1 x 150 mm; Waters Corporation) by an HPLC system coupled with a photo diode array detector [Dittmar, 2008; Ding et al., 2012]. DBC concentrations were calculated from B3CA to B6CA concentrations based on a previously reported algorithm [Dittmar, 2008]. B6CA was excluded from BPCA proportion analysis due to its low signal and low resolution in most samples. Analytical replicates were measured in triplicate with <10% standard deviation. Sample duplicates were collected for 5% of the samples and fell within this <10% standard deviation.

The degree of condensed aromaticity of DBC in each sample was estimated based upon the BPCA distribution [Glaser et al., 1998; Ding et al., 2014b] using the ratio of (B3CA + B4CA)/B5CA. Although a complete understanding of all biogeochemical factors that influence BPCA ratios remains elusive, the ratios can be used to provide preliminary assessments of the environmental dynamics of DBC. For instance, photodegradation can significantly alter BPCA composition, reducing the overall condensed aromaticity of the DBC pool once it reaches surface waters [Stubbins et al., 2012b; Ward et al., 2014]. As such, the DBC signature of samples exposed to solar radiation over long durations may indicate extensive photodegradation, resulting in low signals of B6CA. BPCA ratios can also vary significantly with pyrogenic source material and environmental conditions. For example, higher abundance of BSCA + B6CA, indicating more condensed aromatic DBC, has been associated with DBC in wildfire-impacted watersheds [Wagner et al., 2015], whereas DBC solubilized from urban dust has been shoen to be more enriched in B3CA + B4CA (i.e., less polycondensed BC) [Ding et al., 2014a, 2014b].

Although we understand that the source of pyrolized carbon (i.e., fossil fuels or biomass burning) can influence DBC quality, we expect BPCA compositions for the current sample set to be influenced by the atmospheric residence time of the original aerosol BC source and the degree of exposure to solar radiation (Table 1). Thus, we expected to observe lower (B3CA + B4CA)/BSCA ratios when atmospheric transport time was long and/or opportunities for photodegradation were high.

Table 1. Samples Grouped by Pristine and Nonpristine in Ascending Order of (B3CA + B4CA)/B5CARatio[a]

	(B3CA + B4CA)/B5CA	DBC (μg/L)	DOC (mg/L)	%DOC of DBC	Region	Medium	Geography
			Pristine Snow Sample				
GRIS snow/surface hoar	0.82	1.76	0.67	0.003	GRIS	Snow <1, hoar >5 days	Arctic
Chilean Andes Snow-ET	3.0	1.80	0.83	0.002	Andes	Snow <2 days	Alpine
Niwot Ridge B	3.0	8.93	nd	nd	Rockies	Snow <1 day	Alpine
San Francisco Glacier Snow	3.4	1.44	0.36	0.004	Andes	Snow <2 days	Alpine
WoodFjorden, Svalbard	4.2	1.26	0.54	0.002	Svalbard	Snow <2 days	Arctic
Larsbren	6.3	1.63	0.39	0.004	Svalbard	Snow >5 days	Arctic
Storm Peak Lab	17	9.65	nd	nd	Rockies	Snow >3 days	Alpine
Niwot Ridge A	27	9.27	nd	nd	Rockies	Snow >3 days	Alpine
			Pristine Meltwater Samples				
GRIS supraglacial melt	0.32	12.4	0.14	0.088	GRIS	Supraglacial melt	Arctic
Mer de Glace, Chamonix	1.8	1.51	4.13	0.000	Alps	Glacial stream	Alpine
Bano Morales Stream	1.9	6.12	0.50	0.012	Andes	Glacial stream	Alpine
Lake Joyce 7 m[b]	2.5	2.67	0.50	0.005	Antarctic	Lake	Antarctic
Alatna Pond	3.2	3.60	1.00	0.004	Antarctic	Surface water	Antarctic
Lake Bonney East Lobe 5 m	3.6	1.44	0.40	0.004	Antarctic	Lake	Antarctic
Von Guerrard Stream[b]	3.6	0.89	0.63	0.001	Antarctic	Glacial stream	Antarctic
Ngozumba Glacier, Nepal	3.7	1.05	0.10	0.010	Himalayas	Supraglacial lake	Alpine
Ngozumba Glacier, Nepal	3.8	0.62	0.12	0.005	Himalayas	Supraglacial lake	Alpine
Alaskan Glacial Rivers[c]	4.3	0.53 ± 0.09	0.010 ± 0.004	1.9 ± 0.6	Alaska	Glacial river	Arctic
Lake Bonney West Lobe 5 m[b]	5.4	1.22	0.40	0.003	Antarctic	Lake	Antarctic
Green Lake 4, Colorado	5.4	4.40	nd	nd	Rockies	Lake	Alpine
Ngozumba Glacier, Nepal	6.3	1.54	0.12	0.012	Himalayas	Supraglacial lake	Alpine
Ngozumba Glacier, Nepal	6.7	1.94	0.40	0.005	Himalayas	Supraglacial lake	Alpine
Ngozumba Glacier, Nepal	6.8	1.39	0.83	0.002	Himalayas	Supraglacial Lake	Alpine
			Nonpristine Snow				
Mine 7	3.3	71.4	1.03	0.070	Svalbard	Snow >5 days	Arctic
Upwind of the Mine	7.7	2.36	5.43	0.000	Svalbard	Snow >5 days	Arctic
Next to Coal Power Plant	17	6.84	0.32	0.021	Svalbard	Snow >5 days	Arctic
Upper Snowmobile Track	43	44.4	4.68	0.009	Svalbard	Snow >5 days	Arctic
Lower Snowmobile Track	59	21.6	0.54	0.040	Svalbard	Snow >5 days	Arctic
			Nonpristine Meltwater				
High Park Fire PNF0714[d]	1.3	177	2.97	0.06	Rockies	Terrestrial stream	Alpine
Marble Point	1.5	33.6	4.23	0.008	Antarctic	Surface water	Antarctic
Clear Lake	2.2	47.3	4.30	0.011	Antarctic	Surface water	Antarctic
Lake Fryxell 5 m	2.2	8.28	1.80	0.005	Antarctic	Lake	Antarctic
DLH Pond	3.1	14.0	2.30	0.006	Antarctic	Surface water	Antarctic
Pony Lake	3.4	170	19.0	0.009	Antarctic	Surface water	Antarctic
Longyear River	3.5	4.68	5.83	0.001	Svalbard	Glacial stream	Arctic
Longyearbyen Fjord Water	4.3	20.5	0.33	0.062	Svalbard	Ocean	Arctic

[a]A lower ratio is more indicative of wildfire-derived DBC, and a higher ratio is more indicative of fossil fuel combustion and/or photodegradation.
[b]From *Khan et al.* [2016].
[c]From *Ding et al.* [2014a, 2014b].
[d]From *Wagner et al.* [2015].

The biomass burning smoke aerosol optical thickness (AOT) data used in this study were obtained from the Navy Aerosol Analysis and Prediction System (NAAPS) reanalysis [Lynch et al., 2016]. NAAPS reanalysis is a decade-long global 1 x 1° and 6-hourly 550 nm AOT reanalysis product, which was recently developed and validated at the Naval Research Laboratory. This reanalysis utilizes a modified version of the NAAPS as its core and assimilates quality-controlled retrievals of AOT from Moderate Resolution Imaging Spectroradiometer (MODIS) on Terra and Aqua and the Multiangle Imaging 5 pectroradiometer (MISR) on Terra [Zhang and Reid, 2006; Hyer et al., 2011; Shi et al., 2014]. NAAPS characterizes anthropogenic and bio genic fine aerosol species (including sulfate and primary and secondary organic aerosols), dust, biomass burning smoke, and sea-salt aerosols. Smoke from biomass burning is derived from near-real-time satellite based thermal anomaly data used to construct smoke source functions [Reid et al., 2009], with additional orbital corrections on MODIS-based emissions and regional tunings. The reanalyzed fine- and coarse-mode AOT at 550 nm is shown to have good agreement with the ground-based global-scale Sun photometer network Aerosol Robotic Network ACTs [Holben et al., 1998]. Figure 3 is a series of snapshots at 18Z showing wildfire-related smoke AOT at 550 nm between 18 and 21 June 2014 at 550 nm.

Results and Discussion

As expected, pristine sites had lower median and (2.5, 97.5) quantiles of DBC concentrations (1.8 μg/L (0.62, 12)) than nonpristine sites (21 μg/L (1.6, 170)).ln particular, pristine snow samples exhibited low DBC concentrations (1.8 μg/L (1.3, 10)). The median (B3CA + B4CA)/B5CA ratio for pristine samples from snow was 3 (1, 27) (Table 1 and Figure 11). The remote pristine snow/surface hoar sample from the GRIS had a distinct composition compared to the rest of the data set, containing high levels of BSCA and a ratio of 0.3, below the 2.5% quantile. Similar to the glaciers of Antarctica [Khan et al., 2016], the only potential source of BC to the GRIS is long-range atmospheric transport. The GRIS sample was collected in late June 2014, as wildfires blazed across the Canadian Arctic. As shown by the reanalysis from the NAAPS model, continuous smoke activities occurred over the northwest of Canada between 18 and 21 June 2014 and were transported eastward to the western region of the GRIS (Figure 12). Fresh aerosols from these fires, which appear to have been transported to the western region of the GRIS, would have remained in the surface hoar with limited time for photodegradation prior to sample collection and provide an explanation for the highly condensed aromatic DBC signal observed for this sample. The other most remote pristine snow sample was from Svalbard, which had a DBC concentration of 1.1 μg/L and BPCA ratio of 4. This sample was collected in Woodfjorden, which is several hundred kilometers from the closest town, thus only impacted by atmospheric long-range transport.

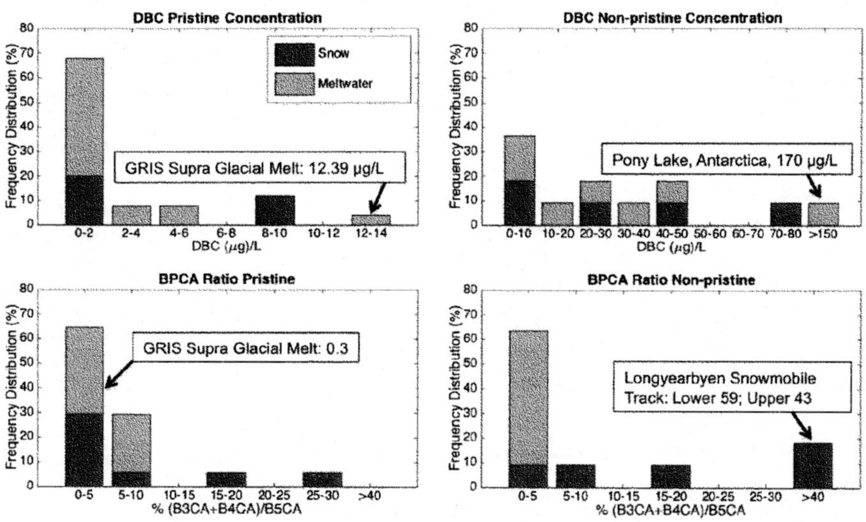

Figure 11. Frequency distribution of pristine and nonpristine snow and meltwater samples from unpublished values in Table 1. Meltwater samples are composed of both snow and ice melt. Note that the maximum x axis value for DBC pristine concentration is 14 μg/L versus 150 μg/L for DBC nonpristine concentration.

Pristine snow samples from the Rocky Mountains had similar DBC concentrations, from 8.9 to 9.7 μg/L, but the (B3CA + B4CA)/B5CA ratios ranged from 3 to 27. Two of the pristine snow samples yielded (B3CA + B4CA)/B5CA ratios higher than the Woodfjorden sample, 4.2. The (B3CA + B4CA)/B5CA ratio was 17 at Storm Peak Lab in Northern Colorado and 27 at Niwot Ridge in central Colorado (Table 1), suggesting that the BC in these remote Colorado samples is less condensed than the remote sample from Svalbard, which may be indicative of more photodegradation of DBC at the Colorado sites prior to sample collection, or a different pyrolized source. These two samples were aged snow samples and exposed to solar radiation longer than the Woodfjorden sample (Table 1). In contrast, a fresh snow sample from a similar location in the Rocky Mountains only yielded a (B3CA + B4CA)/B5CA ratio of 3. The DBC in this fresh snow sample had limited time for photodegradation, which may result in the less condensed BPCA ratio, relative to the aged snow samples from the Rocky Mountains. Again, these sites are >5 km away from known point sources so the enrichment in less condensed DBC may be related to accumulated solar exposure from the time since snow deposition to the time of sample collection and/or the long-range transport of fossil fuel derived soot particles, which have been found on remote glaciers [Stubbins et al., 2012a; Ding et al., 2014a].

Median DBC concentrations were higher in nonpristine snow samples (14 μg/L (1.6, 71)) than pristine snow samples (1.8 μg/L (13, 10)). Furthermore, the median (B3CA + B4CA)/BSCA ratio for nonpristine samples from snow was much higher than pristine snow samples, 12 (3, 59) and 3 (1,27), respectively (Table 1 and Figure 11). The highest DBC concentrations in snow were found in samples collected from Longyearbyen, Svalbard, next to an active coalmine (71 μg/L), a coal burning power plant (6.8 μg/L), and along the primary Longyearbren snowmobile track (22 μg/L) upper track; 44 1-19/L lower track), which receives continuous inputs of fresh anthropogenic BC.

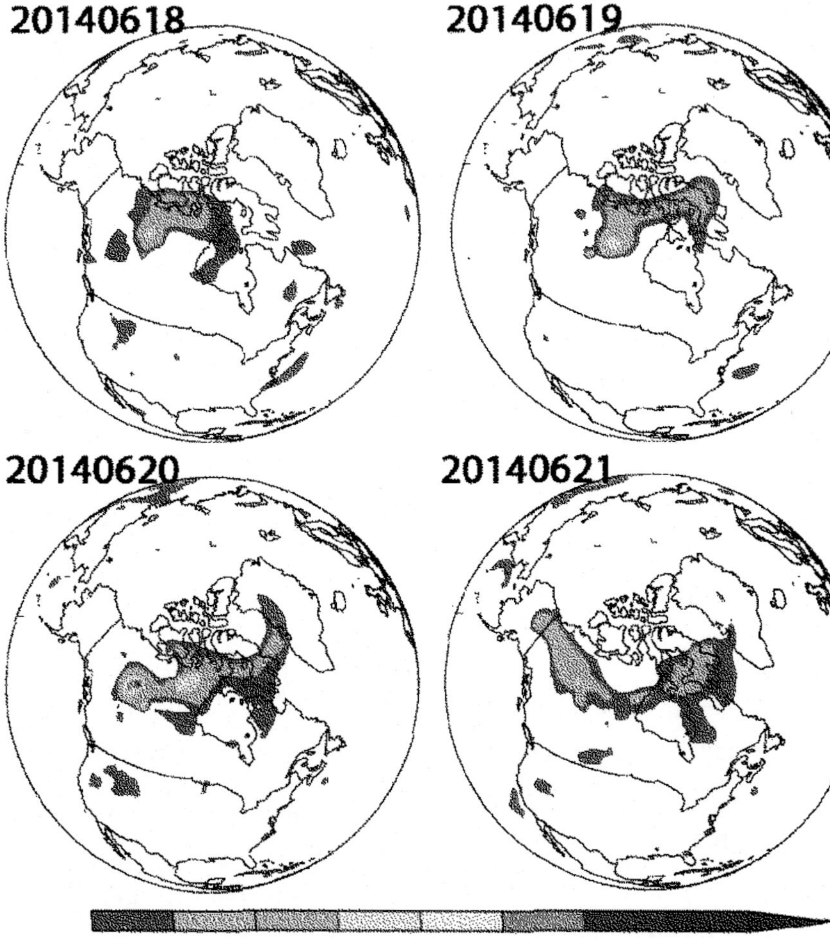

Figure 12. Smoke aerosol optical thickness (AOT) reanalysis at 18Z from the Navy Aerosol Analysis and Prediction System (NAAPS) between 18 and 21 June 2014 shows that wildfire aerosols were transported from the Canadian Arctic to the western region of Greenland Ice Sheet in June 2014. The color bar is atmospheric total column smoke aerosol optical thickness (AOT) at 550 nm, which is unitless.

In contrast to our hypothesis that continuous inputs would result in a signature less indicative to photodegradatlon, these samples also feature the highest (B3CA + B4CA)/B5CA ratios, with the snow next to the coal burning power plant featuring a ratio of 17, and even higher ratios along the snow mobile track (43 upper track; 59 lower track).

Pristine meltwaters had low median DBC concentrations (1.7 μg/L (0.6, 12)), similar to the pristine snow samples (1.8 μg/L {1.3, 10)). The median (B3CA + B4CA)/B5CA ratio for pristine meltwaters, 4 (0.3, 7) (Table 1 and Figure 11), was marginally higher than for pristine snow samples, 3 (1, 27).In contrast to the Nepal meltwaters, the GRIS supraglacial meltwater had a high concentration of DBC (12.4 μg/L) corresponding to the 97.5% quantile. In addition, the GRIS meltwater had a similar low ratio of (B3CA + B4CA)/B5CA to the GRIS snow/surface hoar sample. These BPCA ratios (0.3 and 0.8, respectively) were the lowest in this data set and are lower than the previously reported ratios from terrestrial rivers impacted by wildfires [*Wagner et al.,* 2015]. As noted previously, these low ratios likely reflect fresh and recent BC deposition sourced from Canadian Arctic wildfires (Figure 3). The higher DBC concentration compared to the Nepal supraglacial waters could reflect not only the substantial long range wildfire inputs but also the large surface area of the GRIS, which may provide a larger supraglacial catchment for accumulation and transport of DBC to the ablation area where the sample was collected. Further, the high DBC in the GR!S meltwater provides support to the potential positive feedback proposed by *Keegan et al.* [2014] whereby DBC enriched meltwater is retained as refrozen ice layers in the snowpack and enhances future melting.

Meltwater from supraglacial lakes on the Ngozumba Glacier in Nepal, located at 5000 m, had the lowest DBC concentrations (average of five samples, 1.3 ± 0.49 μg/L) of all meltwater samples. The ratio of these samples was also composed of less condensed DBC, with an average ratio of 5, than supraglacial melt from the GRIS which was composed of highly condensed DBC and had a ratio of 0.3. Although no nearby snow samples were obtained for direct comparison, these low values from the Ngozumba Glacier may be associated with heavy glacial debris coverage in the majority of the ablation region. The debris cover, when thick enough, provides insulation from glacial melt [*Pratap et al.,* 2015] and may reduce supraglacial meltwater generation, along with transport and accumulation of DBC. Although these lakes form an ice cover from about November to March, in the summer they receive large solar radiation inputs, enhanced by the high altitude, which <:auld drive photodegradation of DBC. The less condensed DBC signature could also suggest different combustion sources of DBC between these sampling locations.

Similar to the observations for .snow and ice, nonpristine meltwaters had higher median DBC concentrations, (27 μ/L (4.7, 170)) than pristine meltwaters (1.7 μg/L (0.6,12)). Closed basin nonpristine Antarctic surface waters, such as DLH Pond, Marble Point Pond, Clear Lake, and Pony Lake, contained high concentrations of DBC ranging from 14 to 170, μg/L These lakes are located near active point sources of fossil fuel combustion, such as helicopter flight paths and fueling stations, and exhibited elevated DBC concentrations relative to snow and meltwater. Because these surface water ponds have no outflowing streams, any DBC deposited in the systems will remain and accumulate for millennia, similar to the MDV lakes [*Khan et al.,* 2016], or may be photodegraded. The BPCA ratio of these four nonpristine Antarctic surface waters ranged from 2 to 3. Similarly, the BPCA ratios for Long year River and Advent fjord ocean water in the Norwegian Arctic, both located near the town of Longyearbyen and the local coal burning power plant, were both 4. Such elevated DBC concentrations and BPCA ratios suggest that site proximity to continuous inputs from these particular anthropogenic sources results in high concentrations of less condensed DBC and that long-term photooxidation may not be the only cause of less condensed DBC in the cryosphere.

Conclusions

This study reveals that within the cryosphere, DBC concentrations and composition are widely variable. Regionally, concentration differences exist between remote-pristine sites and those near combustion point sources. Overall, these data support our hypothesis that DBC in the cryosphere that is exposed to longer durations of solar radiation results in less condensed DBC due to photodegradation. Additionally, sites in the cryosphere receiving fresh BC from wildfires, such as on the GRIS where aerosols were deposited directly from Canadian Arctic wildfires, consist of more condensed DBC This signatuie of the DBC from the GRIS was distinct from the sites which receive continuous inputs from local anthropogenic sources of BC, such as along the heavily used snowmobile track in Svalbard, as well as sites which have prolonged sunlight exposure, thus enhancing effects of photodegradation.

While the influence of wildfire inputs throughout the cryosphere may be important regionally, as observed in the samples from the GRIS, the intense solar exposure in high alpine and polar regions, and the characteristics of some anthropogenic combustion sources subject to solar exposure during long-range atmospherics trans port [*Cooke and Wilson,* 1996], may cause the similarity in DBC composition observed throughout the rest of the global cryosphere data set. This includes photodegradation of background levels of wildfire-derived BC, which may also contribute to B3 + 4CA enrichment in the cryosphere [*Stubbins et al.,* 2012b; *Ward et al.,* 2014]. As such, the influence of photodegradation may sometimes be "overwhelmed" by discrete wildfire sources, which likely account for DBC composition of the GRIS samples. Although the current data set allowed for preliminary assessments of contributions to and persistence of DBC in the cryosphere, the observed variability of DBC composition is likely derived from a combination of biogeochemical processes in the environment and different pyrogenic sources.

Previously reported accumulation of millennia old DBC in Antarctic takes [*Khan et al.,* 2016] suggests that DBC is recalcitrant in the cryosphere. The ranges of concentrations observed in this diverse data set of DBC from the global cryosphere indicate that relatively high DBC concentrations persist in both pristine and nonpristine Arctic and Antarctic snow and surface waters. The relatively high concentration of DBC measured in one supraglacial stream on the GRIS may suggest that DBC can be mobilized across ice surfaces. Wildfires are predicted to increase due to climate change [*Flannigan et al.,* 2009], and albedo cannibalism is already influencing meltwater generation on the GRIS (*Tedesco et al.,* 2015). Anticipated longer summer melt seasons as a result of climate change may result in longer durations between snowfalls, enhancing exposure of recalcitrant DBC on snow/ice surfaces, which could further exacerbate surface albedo reduction in the cryosphere.

Acknowledgments

Funding for this work came from the National Science Foundation Graduate Research Fellowship and the National Science Foundation-Graduate Research Opportunities Worldwide (GROW-Chile), award (NSF award: DGE-1144083), the McMurdo Dry Valleys Long-Term Ecological Research Program, award (NSF award: ANT-0423595), the United States Agency for International Development (USAID) Contributions to High Asia Run-off from ice and Snow (CHARIS) project, and the Consortium of Universities for the Advancement of Hydrologic Sciences (CUAHSI) Pathfinder Travel Fellowship. Funding for sample collection on the GRIS came from the Dark Snow Project Additional support for the DBC analyses came from the Florida Coastal Everglades Long-Term Ecological Research Program, award (DEB-1237517) and the George Barley Endowment, This is contribution 835 from the Southeast Environmental Research Center. We thank Eric Sokol for helpful discussion of data analysis and approaches, as well as anonymous reviews for their helpful feedback. The authors also thank Ulyana Horodyskyj, Jason Box, Marek Stibal, and James McPhee for field team support during sampling in Nepal, Greenland, and Chile. The data for this manuscript are presented in Table 1 and the subsequent figures; however, any additional data requests can be directed to alia.khan@colorado.edu. A.L.K. and D.M.M. conceived the study. M.W. and R.A. advised the study design and supported fieid sampling. A.L.K. conducted field sample collection. RJ. provided support for laboratory analysis. A.L.K. and S.W. conducted laboratory analysis, P.X. provided NAAPS model reanalysis. A.L.K.. D.M.M., P.X., S.W., and R.J. contributed to data analysis and writing of the manuscript.

References

Abiven, S., P. Hengartner, M. P. W. Schneider, N. Singh, and M. W. I. Schmidt (2011), Pyrogenic carbon soluble fraction is larger and more aromatic in aged charcoal than in fresh charcoal, *Soil Biol. Biochem.*, *43*(7), 1615–1617, doi:10.1016/j.soilbio.2011.03.027.

Bond, T. C., et al. (2013), Bounding the role of black carbon in the climate system: A scientific assessment, *J. Geophys. Res. Atmos.*, *118*, 5380–5552, doi:10.1002/jgrd.50171.

Cooke, W. F., and J. J. N. Wilson (1996), A global black carbon aerosol model, *J. Geophys. Res.*, *101*(D14), 19,395–19,409, doi:10.1029/96JD00671.

Ding, Y., Y. Yamashita, W. K. Dodds, and R. Jaffé (2012), Dissolved black carbon in grassland streams: Is there an effect of recent fire history?, *Chemosphere*, *90*(10), 2557–2562, doi:10.1016/j.chemosphere.2012.10.098.

Ding, Y., Y. Yamashita, J. Jones, and R. Jaffe (2014a), Dissolved black carbon in boreal forest and glacial rivers of central Alaska: Assessment of biomass burning versus anthropogenic sources, *Biogeochemistry*, *123*, 15–25, doi:10.1007/s10533-014-0050-7.

Ding, Y., K. M. Cawley, C. N. da Cunha, and R. Jaffé (2014b), Environmental dynamics of dissolved black carbon in wetlands, *Biogeochemistry*, *119*, 259–273, doi:10.1007/s10533-014-9964-3.

Dittmar, T. (2008), The molecular level determination of black carbon in marine dissolved organic matter, *Org. Geochem.*, *39*(4), 396–407, doi:10.1016/j.orggeochem.2008.01.015.

Dittmar, T., B. Koch, N. Hertkorn, and G. Kattner (2008), A simple and efficient method for the solid-phase extraction of dissolved organic matter (SPE-DOM) from seawater, *Limnol. Oceanogr. Methods*, *6*, 230–235, doi:10.4319/lom.2008.6.230.

Fellman, J. B., E. Hood, P. A. Raymond, A. Stubbins, and R. G. M. Spencer (2015), Spatial variation in the origin of dissolved organic carbon in snow on the Juneau icefield, Southeast Alaska, *Environ. Sci. Technol.*, 150908131727005, doi:10.1021/acs.est.5b02685.

Flannigan, M., B. Stocks, M. Turetsky, and M. Wotton (2009), Impacts of climate change on fire activity and fire management in the circumboreal forest, *Glob. Chang. Biol.*, *15*(3), 549–560, doi:10.1111/j.1365-2486.2008.01660.x.

Glaser, B., L. Haumaier, G. Guggenberger, and W. Zech (1998), Black carbon in soils: The use of benzenecarboxylic acids as specific markers, *Org. Geochem.*, *29*(4), 811–819, doi:10.1016/S0146-6380(98)00194-6.

Hodson, A. J. (2014), Understanding the dynamics of black carbon and associated contaminants in glacial systems, *Wiley Interdiscip. Rev. Water*, *1*(April), 141–149, doi:10.1002/wat2.1016.

Holben, B. N., et al. (1998), AERONET—A federated instrument network and data archive for aerosol characterization, *Remote Sens. Environ.*, *4257*(98).

Hood, E., J. Fellman, R. G. M. Spencer, P. J. Hernes, R. Edwards, D. D'Amore, and D. Scott (2009), Glaciers as a source of ancient and labile organic matter to the marine environment, *Nature*, *462*(7276), 1044–1047, doi:10.1038/nature08580.

Hood, E., T. J. Battin, J. Fellman, S. O'Neel, and R. G. M. Spencer (2015), Storage and release of organic carbon from glaciers and ice sheets, *Nat. Geosci.*, *8*(2), 1–6, doi:10.1038/ngeo2331.

Hyer, E. J., J. S. Reid, and J. Zhang (2011), An over-land aerosol optical depth data set for data assimilation by filtering, correction, and aggregation of MODIS Collection 5 optical depth retrievals, *Atmos. Meas. Tech.*, *4*, 379–408, doi:10.5194/amt-4-379-2011.

Keegan, K. M., M. R. Albert, J. R. Mcconnell, and I. Baker (2014), Climate change and forest fires synergistically drive widespread melt events of the Greenland Ice Sheet, *Proc. Natl. Acad. Sci. U. S. A.*, *111*, 7964–7967, doi:10.1073/pnas.1405397111.

Khan, A. L., R. Jaffé, Y. Ding, and D. M. McKnight (2016), Dissolved black carbon in Antarctic lakes: Chemical signatures of past and present sources, *Geophys. Res. Lett.*, *43*, 5750–5757, doi:10.1002/2016GL068609.

Lynch, P., et al. (2016), An 11-year global gridded aerosol optical thickness reanalysis (v1. 0) for atmospheric and climate sciences, *Geosci. Model Dev.*, *9*, 1489–1522, doi:10.5194/gmd-9-1489-2016.

Mcbeath, A. V., R. J. Smernik, and E. S. Krull (2013a), Organic geochemistry: A demonstration of the high variability of chars produced from wood in bushfires, *Org. Geochem.*, *55*, 38–44, doi:10.1016/j.orggeochem.2012.11.006.

Mcbeath, A. V., R. J. Smernik, E. S. Krull, and J. Lehmann (2013b), ScienceDirect: The influence of feedstock and production temperature on biochar carbon chemistry—A solid-state ^{13}C NMR study, *Biomass Bioenergy*, *60*, 121–129, doi:10.1016/j.biombioe.2013.11.002.

Pratap, B., D. P. Dobhal, M. Mehta, and R. Bhambri (2015), Influence of debris cover and altitude on glacier surface melting: A case study on Dokriani Glacier, central Himalaya, India, *Ann. Glaciol.*, *56*(70), 9–16, doi:10.3189/2015AoG70A971.

Reid, J. S., et al. (2009), Global monitoring and forecasting of biomass-burning smoke: Description of and lessons from the Fire Locating and Modeling of Burning Emissions (FLAMBE) program, *IEEE J. Sel. Top. Appl. Earth Obs. Remote Sens.*, *2*(3), 144–162.

Schneider, M. P. W., M. Hilf, U. F. Vogt, and M. W. I. Schmidt (2010), The benzene polycarboxylic acid (BPCA) pattern of wood pyrolyzed between 200°C and 1000°C, *Org. Geochem.*, *41*(10), 1082–1088, doi:10.1016/j.orggeochem.2010.07.001.

Shi, Y., J. Zhang, J. S. Reid, B. Liu, and E. J. Hyer (2014), Critical evaluation of cloud contamination in the MISR aerosol products using MODIS cloud mask products, *Atmos. Meas. Tech. Discuss.*, 1791–1801, doi:10.5194/amt-7-1791-2014.

Stubbins, A., et al. (2012a), Anthropogenic aerosols as a source of ancient dissolved organic matter in glaciers, *Nat. Geosci.*, *5*(3), 198–201, doi:10.1038/ngeo1403.

Stubbins, A., J. Niggemann, and T. Dittmar (2012b), Photo-lability of deep ocean dissolved black carbon, *Biogeosciences*, *9*(2002), 1661–1670, doi:10.5194/bg-9-1661-2012.

Tedesco, M., S. Doherty, X. Fettweis, P. Alexander, J. Jeyaratnam, E. Noble, and J. Stroeve (2015), The darkening of the Greenland ice sheet: Trends, drivers and projections (1981–2100), *Cryosphere Discuss.*, *9*(5), 5595–5645, doi:10.5194/tcd-9-5595-2015.

Wagner, S., K. M. Cawley, F. L. Rosario-Ortiz, and R. Jaffé (2015), In-stream sources and links between particulate and dissolved black carbon following a wildfire, *Biogeochemistry*, *124*, 145–161, doi:10.1007/s10533-015-0088-1.

Ward, C. P., R. L. Sleighter, P. G. Hatcher, and R. M. Cory (2014), Insights into the complete and partial photooxidation of black carbon in surface waters, *Environ. Sci.: Processes Impacts*, *16*(4), 721–731, doi:10.1039/c3em00597f.

Wiedenmeyer, M., et al. (2015), Aromaticity and degree of aromatic condensation of char, *Org. Geochem.*, *78*, 135–143, doi:10.1016/j.orggeochem.2014.10.002.

Zhang, J., and J. S. Reid (2006), MODIS aerosol product analysis for data assimilation: Assessment of over-ocean level 2 aerosol optical thickness retrievals, *J. Geophys. Res.*, *111*, D22207, doi:10.1029/2005JD006898.

Ziolkowski, L. A., and E. R. M. Druffel (2010), Aged black carbon identified in marine dissolved organic carbon, *Geophys. Res. Lett.*, *37*, L16601, doi:10.1029/2010GL043963.

INDEX

A

B

C

F

G

S

T

U

V

W

Y